Mt. Lebanon Public Library
16 Castle Shannon Blvd
Pittsburgh, PA 15228-2252
412-531-1912 | www.mtlebanonlibrary.org

THE SECRET LIFE OF SECRETS

THE SECRET LIFE OF SECRETS

HOW OUR INNER WORLDS SHAPE WELL-BEING,
RELATIONSHIPS, AND WHO WE ARE

Michael Slepian

CROWN
NEW YORK

Published in the United States by Crown, an imprint of Random House, a
division of Penguin Random House LLC, New York.

CROWN and the Crown colophon are registered trademarks of Penguin
Random House LLC.

Library of Congress Cataloging-in-Publication Data
Names: Slepian, Michael, author.
Title: The secret life of secrets: how they shape our relationships, our
    well-being, and who we are / Michael Slepian.
Description: First edition. | New York: Crown, [2022] | Includes
    bibliographical references and index.
Identifiers: LCCN 2021054113 (print) | LCCN 2021054114 (ebook) |
    ISBN 9780593237212 (hardcover) | ISBN 9780593237236 (ebook)
Subjects: LCSH: Secrecy. | Family secrets. | Identity (Psychology) |
    Interpersonal relations.
Classification: LCC BJ1429.5 .S54 2022 (print) | LCC BJ1429.5 (ebook) |
    DDC 302.2—dc23/eng/20211222
LC record available at https://lccn.loc.gov/2021054113
LC ebook record available at https://lccn.loc.gov/2021054114

International edition ISBN 978-0-593-44354-5

Printed in the United States of America on acid-free paper

crownpublishing.com

9 8 7 6 5 4 3 2 1

First Edition

Book design by Caroline Cunningham

For Rachel

CONTENTS

Preface: A Secret Revealed   ix

1. What Is a Secret?   3

2. The Birth of Secrets   26

3. Secrets on the Mind   51

4. The Three Dimensions of Secrets   74

5. Concealing Our Secrets   105

6. Confessing and Confiding   129

7. Positive Secrets   157

8. Culture and Coping   173

9. Secrets Shared   189

Acknowledgments   195

Notes   199

List of Illustrations   223

Index   225

PREFACE

A Secret Revealed

I watched nervously as snow and ice closed down other East Coast destinations. Newsclips showed even D.C. had been brought to a standstill, but flights were still landing at New York City airports for the time being. I was lucky in the end; my flight from California made it off the ground on time, and I arrived in Manhattan without issue.

I had seen the Columbia campus only once before, as a high school student visiting the one college on my application list that I had zero chance of getting into, and now here I was, standing at the podium in the front of one of its lecture halls. My worry about the winter storms was replaced by a new anxiety. I was about to deliver my "job talk," a ninety-minute research lecture and discussion that covered a range of topics, from how we judge others' trustworthiness to how we judge the steepness of hills. I began the talk with my latest research on secrets.

The standard job interview in academia is an exhausting series of back-to-back meetings *that you've spent countless hours studying for*—interrupted only by the most important talk you will ever give, *for which you've spent months preparing and practicing*, and a handful of bathroom breaks, should folks remember to offer them. This interview was no different. When I began, hands went up right away, and the people raising them had questions ranging from the fundamental (what is a secret?) to the formidable (what about culture?).

But this isn't a story about my job interview. This day is forever seared in my memory for a different reason. Besides the talk, I remember two other things: one, after a full day of meetings, my future colleagues took me out to dinner, and then some of us continued on to my hotel bar; and two, later that same night, the very research I had spent that day showcasing would take an unexpected and dramatically personal turn.

It was sometime past midnight when we ordered another round of drinks, and when I realized the night felt like one spent with good friends. The interview already felt like it was from a time long ago, a distant memory. I suddenly remembered that my phone had vibrated earlier, and so I peeked to see why. A missed call from my dad. This was highly unusual. My dad never calls out of the blue, let alone close to midnight.

Thirty minutes later, we ordered the check instead of another round, and I saw a second missed call from my dad. My mind could only conjure the worst scenarios. Surely there had been a death in the family, or some other tragedy. Whatever he wanted to talk about, it seemed like it couldn't wait.

I returned my dad's phone call. "I need to share something with you," he said. "Can you sit down?" He then told me a story about my parents' initial attempts to have children, and how they failed. Time after time, they failed, and eventually they learned why. My father is biologically unable to have children.

My father was telling me that he wasn't my biological father: that I was conceived by artificial insemination from an anonymous donor. This secret was being revealed to me at the conclusion of one of the most important days of my professional life. I was not sitting down.

I felt light-headed. A rush of adrenaline returned in full force as if I were back at the podium giving my talk, except now I was the one with questions. My dad answered my first big one before I could even ask it. My brother, born five years later, is in fact my half-brother, conceived from a different donor.

I patiently listened until my father finished his bombshell of a story. Then it was my turn. I reassured him that what he told me did not change anything between us. He's my father, as he always has been. But I had questions. Why was this secret kept from me? Why was he telling me about it, and why now? Who else knew? It turned out that all my grandparents, uncles, and aunts—the entire family apart from my brother and me—had known the whole time. How in the world did everyone manage to keep this secret for so long, and what was it like to do so?

I've spent the past ten years conducting research that speaks to the very questions I asked my dad that night. It would later dawn on me that my century-old field had yet to establish a psychology of secrecy precisely because it hadn't been asking questions like these. Psychologists have long been interested in how people form relationships and connect with each other; but why we hold back from others, and the consequences of not letting other people into our inner worlds, has largely been overlooked.

Prior researchers assumed that hiding information during conversations was the whole of secrecy, and they designed clever experiments based on that assumption. But when we look at real secrets such as the ones you are currently keeping, and when we ask how they affect you from day to day, a different picture begins to emerge.

More than a story of sneakiness or deceit, the story of secrecy is one of the inner workings of the mind, our sense of self and our relationships, how we cope with life's challenges, and what makes us human.

All people have secrets, but not all secrets are alike. Some secrets don't hurt, and others do. But which secrets hurt, and why? Psychologists had yet to ask this question the night I learned my family's secret, but in my research I've asked it more than a hundred times since. Thanks to tens of thousands of research participants who have shared their secrets with me, we can now shine a light on the secret life of secrets, and reveal the multiple ways in which they pervade our lives and what we can do to better live alongside them.

We may not want our secrets to be known, but we do want our selves to be known. Navigating this tension is key to our relationships with others. One of the core reasons that secrets are so difficult to keep is that secrecy goes against our human impulse to share our experiences with others. Indeed, not all secrets remain secrets. Sometimes we confess. Other times we confide.

Whether or not your secrets remain hidden, you will come away from this book with a better understanding of why we keep secrets, how they affect us in ways that you might not be aware of, and how to thrive while carrying them.

THE SECRET LIFE OF SECRETS

CHAPTER 1

What Is a Secret?

The very first scene of *The Sopranos*, the HBO show that set the course for a new age of television, begins with a secret. Tony Soprano is sitting in a waiting room and he is looking around the room. It's quiet enough to hear a clock ticking, which makes the sound of a suddenly opened door startling. Dr. Melfi calls Tony into her office, he follows her in, and they both take a seat. Tony has the posture of somebody who is trying to look relaxed more than someone who is truly at ease. He looks to Dr. Melfi expectantly. She returns his gaze. The silence is awkward and Tony taps his fingers to fill it. He breaks eye contact, glances up and away and then looks back into her eyes, and breathes in deeply.

Tony Soprano is seeing a therapist. He doesn't want a soul to know. If any of his pals and business associates were to learn of it, they would think less of him; he is sure of it. It would make him look soft and weak; hardly what one would want to project as the head of an organized crime ring. Tony knows there is little risk of his secret being discovered. After all, in his line of work, secrecy is a job requirement; it's even central to the oath of Omertà, the code of silence that criminal organizations have upheld for centuries. And it's not like "Are you seeing a therapist?" is a question that comes up in everyday conversation. Yet the secret weighs heavily on him. Why was this secret so burdensome for Tony?

Psychologists have long believed that secrets take a toll on our mental and physical health, but the question of *why* has proved difficult to answer. For years, most researchers assumed that the act of *hiding* our secrets was what made secrets harmful to our health. The classic study design placed research participants in conversation with another person, from whom they were asked to conceal a secret. Often, the other person in the study was instructed to ask questions about the very secret the participant was instructed to hide. But does this situation capture the full scope of how people experience their secrets?

DEFINING SECRECY

"Nothing is harder than living with a secret that can't be spoken." Edward Snowden wasn't worried about people asking about his secret, but rather was worried about how he could safely get his secret out. Snowden discovered that the National Security Agency was secretly engaging in mass global surveillance. "They could just spy on the world without telling a soul," he wrote in his autobiography. Snowden believed that the program undermined the very purpose of the NSA, which was to protect civil liberties, not violate them.

Snowden decided to become a whistleblower, but this presented two problems. First, there was the scope and complexity of the surveillance system. For the information to be credible, Snowden needed to reveal how the surveillance program worked. "The way to reveal a secret program might have been merely to describe its existence, but the way to reveal programmatic secrecy was to describe its workings," he later wrote. "This required documents, the agency's actual files—as many as necessary to expose the scope of the abuse," which brings us to the second problem. The U.S. government does not take kindly toward the illegal leaking of classified information. "I knew that disclosing even one PDF would be enough to earn me prison."

Every move had to go undetected. While the NSA computers in use sent data processing and storage to the cloud, leaving digital paper trails, Snowden realized that the old NSA computers no longer in use, just sitting in an office discarded, could not be so easily tracked. But it would look strange if anyone saw him using those ancient computers, and so he copied the critical files at odd hours under the cover of darkness. "I'd be sweating, seeing shadows and hearing footsteps around every corner." He downloaded files to a tiny memory card—one too small to set off any metal detectors—which he carried out of the building hidden in his pocket and, on one occasion, inside his Rubik's Cube. "In other attempts, I carried a card in my sock, or, at my most paranoid, in my cheek, so I could swallow it if I had to."

To leave no digital trace that could be connected back to him, Snowden would drive around Oahu, Hawaii, where he was stationed, intercepting Wi-Fi connections and sending the files off, using a different network each time. "I contacted the journalists under a variety of identities, disposable masks worn for a time and then discarded . . . You can't really appreciate how hard it is to stay anonymous online until you've tried to operate as if your life depended on it."

Snowden spent more than six months documenting the secret surveillance program and surreptitiously sharing it with journalists. After the final files were sent, the next stage of his plan was to flee the country. Snowden could not even tell his girlfriend of his plan. The stakes were too high. "Not wishing to cause her any more harm than I was already resigned to causing, I kept silent, and in my silence I was alone."

On the same day his girlfriend, now wife, left for a camping trip, Snowden fled to Hong Kong, where he waited for two journalists to arrive. Days later, the world would learn about the NSA's mass global surveillance and a photo of Snowden would be plastered across every news program.

Snowden described the experience of keeping his secret as isolating, and spoke of the frustration of not having someone to talk to. "Hadn't I gotten used to being alone, after all those years spent hushed and spellbound in front of a screen? . . . But I was human, too, and the lack of companionship was hard. Each day was haunted by struggle, as I tried and failed to reconcile the moral and the legal, my duties and my desires." Snowden also made a distinction between the classified information that he and his coworkers were together keeping secret, and his own secret: he was going to blow the whistle. "At least you're part of a team: though your work may be secret, it's a shared secret, and therefore a shared burden. There is misery but also laughter. When you have a real secret, though, that you can't share with anyone, even the laughter is a lie. I could talk about my concerns, but never about where they were leading me."

You might not be a New Jersey mob boss, nor is it likely that you've ever discovered a top-secret government program, but Tony Soprano's and Edward Snowden's struggles with secrets may feel familiar all the same. Beyond the drama of organized crime, Tony struggled with anxiety and depression. And aside from the international intrigue of a massive spying program, Snowden's fundamental struggle with his secret was that of feeling isolated and alone. Despite their ability to successfully hide their secrets in conversation, Tony Soprano and Edward Snowden felt burdened by their secrets.

Being asked a question about your secret might be the most awkward experience you can imagine, but how common of an experience is this, really? I've never outright asked any of my friends, even my best friends, if they ever cheated on their partner, had an abortion, were abused during their childhood, and so on. Dodging questions about our secrets doesn't happen in real life as often as it does in our worst fears, or in psychology experiments for that matter. In fact, that kind of situation captures only a small slice of the experience of secrecy.

So then, what *is* a secret? Secrets are not physical things out in the world. You can't put one under a microscope, and there is no region of the brain where all our secrets sit. And while hiding information is an action that you may take to keep a secret, this action is not the secret itself. The problem with defining secrecy by the actions we take to hide secrets is that we don't have to hide them very often, and hiding them is not necessarily difficult. When asked where he was earlier in the day, for example, Tony Soprano could easily come up with any number of responses instead of answering truthfully, "I was at therapy." And plenty of secrets don't require upkeep or lies to maintain. This is why we must define secrecy not as something we do, but as an intention: *I intend for people to not learn this thing.*

When we recognize that dodging questions and biting our tongues in conversation is only a single chapter in the story of secrecy, a much broader understanding of secrets—and the roles they play in our lives—can come into focus.

SCALING HILLS

At our holiday party one year, with the living room lights on low, my wife and I put on a video of a fireplace with a burning log to heighten the cozy vibe. The internet is replete with options for those seeking such a video, and the one we chose was several hours long, even featuring realistic crackling sounds. Several partygoers independently remarked that they could almost feel the warmth of the fire emanating from the TV. This wasn't because our TV was overheating or that our friends were drunk and confused. The experience of being washed in orange and yellow light with crackling sounds is something we very much associate with the *feeling* of physical warmth, such as when in front of a real fireplace or campfire.

Neuroimaging studies find that imagining a sensation activates the same neural regions associated with experiencing the sensation. This explains not only the fireplace illusion, but also why it is easier to picture an image with your eyes closed rather than open. It is distracting to look at something when trying to visually imagine something else, because the brain regions involved in both processes are the same.

My original interest in secrecy was in this vein. In a manner similar to the fireplace illusion (where partygoers had the uncanny perception of warmth that wasn't actually there), I wondered: Could thinking about a secret lead to a perception like that of carrying physical weight, making other tasks seem more effortful and challenging?

People talk about secrets in a curious way, often in terms of carrying them around and being weighed down by them. Interested in this physical weight metaphor for being *burdened*, I ran a study in which I randomly assigned a group of online research participants to do one of two things. I asked one group to think about a "big" secret they were keeping, and another group of participants to think about a "small" one. Then we showed a series of images, asking participants to provide their best estimate of what they saw. In one, participants saw a lush green park and were asked what temperature it appeared to be outside. But the real number we cared about was their answer to the next question. We showed the participants an image of a grassy hill and asked them to estimate the steepness of the incline in degrees.

When people feel fatigued, they judge the world around them as more challenging and forbidding, judging hills as steeper and distances as farther. This is not a bug but a feature of the human perceptual system. On a hike, you wouldn't want to be cutting it close, having just barely enough energy to get to the top. Overestimating the steepness of a hill—and therefore the work required to scale it—prevents us from embarking on overly ambitious climbs. And this is just what our research participants, like the many who came before them, did. They overestimated the hill's slant (judging the approximately 25-degree hill, shown face-on, to be about 40 degrees steep). But the participants who thought about *big* secrets judged the hill as even steeper. They perceived the world around them as if they were carrying a greater burden.

Before you take out your protractor, know that these perceived hill slants are interesting only to the extent that they capture feeling like the external world poses challenges. In another study, we directly measured this. We recruited people who were currently in a relationship, and after asking some questions about their relationship, we hit them with a big one: Have you ever been unfaithful? If they said yes, we then asked them how preoccupied they were with the infidelity, specifically how much they thought about it and how much it bothered them. And then we asked them how effortful it would be to perform a variety of tasks: carrying groceries upstairs, walking a dog, and helping someone move. The more preoccupied the participants were with their infidelity, the more effortful they found those other tasks.

When partygoers felt the heat of a fictional fire, their brain's perceptual system was responsible: it was whispering *warmth*. And here, when participants were preoccupied by their secret infidelities, and other tasks seemed more effortful, again their perceptual system was responsible: it was whispering *burdened*.

It's likely I would have stopped studying secrecy at this

point had it not been for one thing. When another research team repeated the hill study, the results didn't replicate; they did not consistently find a difference in judged hill slants after thinking about "big" versus "small" secrets. And so, I went back to the lab to determine what went wrong. Suddenly, I saw an anonymous reviewer's question on the research in a new light: *When asking participants to think about a secret—and varying just one word across the two requests—what did the researchers mean by a "big" versus a "small" secret?* The reviewer added: *"big secrets" don't necessarily have to be "heavy secrets."*

Years later, I would prove that reviewer right. When we ran our original study again, we asked the participants how preoccupied they were with their secret (as we had of the participants who committed infidelity). The participants who thought of "big" secrets reported being somewhat more preoccupied by their secrets, relative to the participants who thought of "small" secrets, but it turned out that the "size" of the secret wasn't what mattered. Rather, the more preoccupied participants were by their secret, the steeper they judged the hill to be.

Thinking back to the secret I revealed in the Preface, by most standards, "I am not your biological father" is a big secret. But having to carry this secret was not always burdensome for my parents. Much of the time, the secret was far from their minds. Other times not so much, such as when people would remark that my brother took after my father whereas I took after my mother, making the secret impossible for my parents to ignore. Which child looks more like which parent is a typical conversation with new parents, and so the secret was more burdensome during those early years.

But that burden would fade over time, they told me, until something would occasionally happen to bring the secret back to the top of their minds. It was always the same "big" secret, but it wasn't always preoccupying. This distinction would prove important.

We ran the hill slant study again, and specifically asked one group of participants to think about a secret that preoccupies them. As we suspected, those participants more consistently judged the hill as steeper than those asked to think about a non-preoccupying secret. Taken together, these studies suggested that even if a secret seems "big," it is not necessarily burdensome to carry, and the secrets that occupy our thoughts are the ones that come with the most mental weight.

The more frequently participants had their secrets on their minds, the more thinking about those secrets was associated with a sense of burden.

But nobody was hiding anything at any point during these studies. There was no second person asking questions, let alone those designed to make it difficult for the participants to keep their secret. And yet, just like Tony Soprano and Edward Snowden, my participants felt burdened by their secrets when nobody was asking about them.

THE SECRETS WE KEEP

Take a moment to consider the following experiences. For each one, ask yourself: Have I had an experience like that? And if so, was it ever a secret? Even if you have discussed the experience with someone, if you still intend to keep it from someone else, then it would still count as a secret.

- Hurt another person (emotionally or physically).
- Illegal drug use, or abuse of a legal drug (e.g., alcohol, painkillers).
- Habit or addiction (but not involving drugs).
- Theft (any kind of taking without asking).
- Something illegal (other than drugs or theft).
- Physical self-harm.
- Abortion.
- A traumatic experience (other than the above).
- A lie.
- A violation of someone's trust (other than by a lie).
- Romantic desire (while single).
- Romantic discontent (being unhappy in a relationship).
- Extra-relational thoughts (thoughts about having relations with another person while in a relationship).
- Emotional infidelity (having an inappropriate emotional connection with someone, engaging in something intimate other than sex).
- Sexual infidelity.
- A relationship with someone who is cheating on someone else to be with you.
- Social discontent (unhappy with a friend, or unhappy with current social life).
- Physical discontent (dislike of appearance or something physical about yourself).
- Mental health struggles.

- Inappropriate behavior at work or school (or lying to get hired or accepted).
- Poor performance at work or school.
- Profession/work discontent (unhappy with your situation at work or school).
- A planned marriage proposal.
- A planned surprise for someone (other than a marriage proposal).
- A hidden hobby or possession.
- A hidden current (or past) relationship.
- A family secret.
- Pregnancy.
- Sexual orientation or gender identity.
- Sexual behavior (other than sexual orientation).
- Not having sex.
- A hidden preference (or non-preference) for something.
- A hidden belief (e.g., political, religious, views about social groups, prejudices).
- Finances (e.g., spending, amount of money you have).
- A hidden current (or past) employment or school activity.
- An ambition, plan, or goal for yourself.
- Unusual or counternormative behavior (unrelated to the above).
- A specific story you keep secret (unrelated to the above).

These are the most common types of secrets that people keep—38 in total. I should mention that these categories come from a survey I conducted of 2,000 people living in the

United States that simply asked: *What is a secret that you are currently keeping?*

There were a few distinctions that our participants cared about, which we respected when making the list. For example, the difference between so-called emotional infidelity and sexual infidelity was clearly important to the people with the secret, and so the list makes this distinction. Likewise, illegal drug use could have been lumped with other illegal activity, but our participants were telling us that drug use was the more pertinent issue at hand, not whether it was legal or not. And theft, which we broadly define as taking without asking, was often seen as its own category: maybe you stole something as a child once, or you "borrowed" something knowing that you would never return it (I'll admit here to both, helping myself to my brother's Halloween candy and a friend's pair of socks). Many of these secrets deal with relationships and sex, a theme we'll see again and again. Others deal with our ambitions, our careers, our finances. Still others lay bare our shames and embarrassments.

When my colleagues and I use this list in our studies, we provide examples of what we mean by "a violation of someone's trust" (snooping on someone, revealing others' information without their knowledge or consent, breaking or losing something that belongs to someone without telling). Likewise, for "romantic desire," we provide a few examples so that our participants know what we have in mind (having a crush on someone, being in love with someone, wanting relations with a specific person). For the mental health category, we also give examples (fears, anxieties, depression, mental disorders, eating disorders). And likewise, we

offer examples for sexual behavior: pornography, mastur-
bation, fantasies, kinks, etc.—letting that "etc." stand in for
the cornucopia of sexual proclivities that one might keep
secret.

I've shared this list with more than 50,000 research participants living across the United States and across the
globe. On average, they indicate having had 21 of the 38 ex-

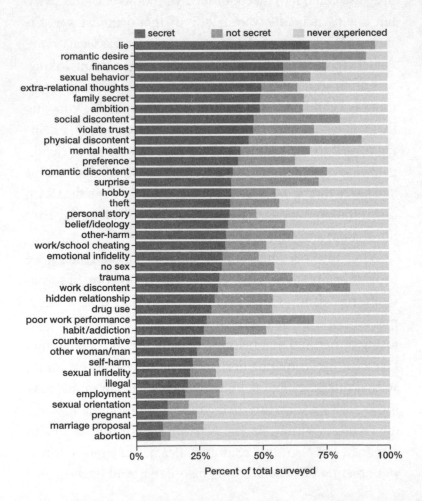

periences, 13 of which they keep secret. And 97% of my participants say they currently have at least one of the secrets from the list. (These numbers primarily reflect my American participants. We'll look at my global sample when we consider the role of culture in Chapter 8, but for better and for worse, I find that secrets affect us in similar ways, wherever we're from.)

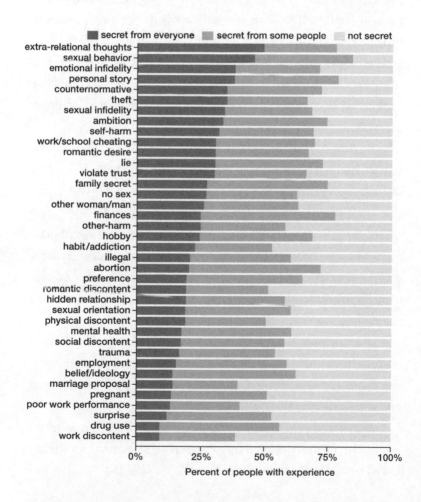

BY THE NUMBERS

Among more than 50,000 research participants I've surveyed, the most common secrets include a lie we've told (69%), romantic desire (61%), sex (58%), and finances (58%).

Next, about half of my participants report having a family secret, a secret ambition, or secret "extra-relational thoughts" (defined as having thoughts about relations with another person while already in a relationship). Former president Jimmy Carter once shocked the nation with colorful language, admitting to having this very secret in an interview with *Playboy* during his presidential campaign, saying, "I've looked on a lot of women with lust. I've committed adultery in my heart many times."

Of course, you can't keep an experience secret unless you've had that experience, which is why we have to look at *not just the overall percentage* of people who keep a given secret (the total count divided by the number surveyed—the graph on the left). We also need to examine what percentage of people *with the experience* keep it secret (the count of people with a given experience who keep it secret *divided by the total number of people who've ever had the experience, secret or not*—the graph on the right).

For example, 36% of all my participants say they have a secret about cheating at work or school. *Among those who have ever cheated at work or school,* however, most keep it secret from at least some people (69%).

Having a secret about extra-relational thoughts is more common (50%) than having a secret emotional infidelity (34%) or a secret sexual infidelity (21%). This is primarily because, whether they keep it secret or not, more people have the first experience (65%), than either of the latter two (48% and 31%, respectively). But also, among those with the experience, extra-relational thoughts are more frequently kept secret (77%), compared to emotional infidelity (71%) and sexual infidelity (68%).

The least commonly kept secrets include sexual orientation (12%), pregnancy (12%), a marriage proposal (10%), and abortion (9%). Relative to all the other secrets on the list, fewer people in my sample *currently* have these secrets.

Abortion is the least common experience on the list: 17% of the women in my data report having had an abortion, which is consistent with the national average. *Among those with this experience,* however, most keep it a secret from at least some people (72%).

Being pregnant or planning to propose marriage to your partner are less common secrets as well, but for a different reason. These secrets (along with other surprises) can only be kept for so long, so fewer people have these secrets at any given time. These more positive secrets are also kept for different reasons than many other types of secrets, something we'll talk about in Chapter 7.

While all secrets share an element of intention, that intention need not apply to all situations or people. Some secrets you keep entirely to yourself, and others you've shared with at least one person but still keep from others. And so, you could have many secrets, but very few that you keep entirely to yourself. On average, I find people concurrently have five secrets they've never told anyone about (a "complete secret") and eight more secrets they have discussed with at least one person but still intend to keep from others (a "confided se-cret"). This is where the average total of thirteen secrets comes from.

Some secrets are more frequently confided than others. For example, among people who say they have a detail about finances that one might keep secret, 53% keep it secret but have also confided in someone, and 24% keep it entirely to themselves (the remainder do not keep it secret). Compare these numbers to those around "extra-relational thoughts." Among people who've had romantic or sexual thoughts about another person outside of their relationship, only 28% confide in someone about having this secret, and 49% keep it entirely to themselves. Confiding in others usually brings benefits, but not always. We'll discuss confessing and confiding in Chapter 6.

When I say that people on average have *thirteen secrets from the list* at any given time, you should take that to be an under-estimate, as what this count really reflects is the number of categories of secrets people have. You could have two or more secrets that fall into the same category, especially for those that are broad, such as finances, sexual behavior, violating someone's trust, or doing something illegal.

You might be wondering how else these secrets compare to each other. Are some more harmful than others? We can't answer that question just yet. To do so, we need some way to compare the secrets to each other, some ruler to hold them up against. We'll talk about that in Chapter 4.

The list of secrets I've shared with you won't include every single secret you have, as secrets can be about anything. But the list is fairly comprehensive. When I ask people to simply tell me about a secret they are keeping, 92% of the time it fits into one of the 38 categories from our list. This, as we will see in the coming chapters, means that we are not so alone in the secrets we keep, despite how isolating the experience of secrecy can feel. Far from being what makes us different from others, secrecy is what we have in common.

IS IT A SECRET?

When you go through the 38 categories, there may be a few for which you think, *Well, I have had this experience, and I don't really tell people about it, but is it a secret?*

What sets a secret apart from other things we don't talk about is an intention—specifically, the *intention* to keep the information unknown. To help identify which experiences count as a secret, we need to distinguish secrecy from privacy.

You can draw a line between secrecy and privacy by considering secrecy as an intention to hold specific information back, and privacy as a reflection of how much you broadcast personal information, in general. People who are more private require closeness before they let you in. Yet those who are less private may be happy to disclose personal information, and not just to friends and family, but to coworkers, acquaintances, and even people they've just met. You may not want to discuss your sexual experiences at work out of concern for privacy (and for what is appropriate), but this is very different from wanting to keep some specific experience a secret. In both cases, you are taking control of your personal information, but for different reasons.

Aside from sex, money is another example of something you may not talk about but may not be intentionally keeping secret. You might not talk about your paycheck out of concern for privacy, rather than wanting *nobody* to ever know what it looks like. At the same time, there may be other specifics you intend to keep hidden, such as a particularly unwise financial decision. These examples help us see that privacy and secrecy can coexist, and there can be gray area in between. So, can we ever really separate them? Yes, and the person who knows best—whether something is private or secret—is you.

I find in my research that the more immoral we consider a personal experience or action, the more it feels like a secret, rather than something that is merely private. I also find that the more we think others would find the information relevant to their own lives, the more something unsaid feels like secrecy instead of privacy.

We know this from a study involving 1,000 participants in committed relationships. I asked the participants to think about something they had not disclosed to their romantic partner. This was easy for them to do. We all have many such things, ranging from the consequential to the mundane. Some of the things people hadn't disclosed were acts they considered highly immoral, like cheating on their partner and misrepresenting their past. The participants said that these felt very much like secrets. But other things did not seem immoral. For example, one participant told me he quite enjoys having the apartment to himself, and doesn't mind when his partner is away for the weekend. In fact, it makes him quite happy. Another participant told me that her partner doesn't know how much she spends on yarn. These things didn't feel like they mattered all that much, and so not mentioning them didn't feel like keeping secrets.

A commonly avoided topic within romantic relationships is information about past relationships. Sure, when we first get together with someone there is some pertinent information to trade, but we tend not to see much value in discussing the intimate facts with our current partner. It's not that this information is being held back, but rather we have no need to talk about it.

I find in my research that there is another major reason you may choose to avoid a conversation topic: you are trying to avoid a conflict. A conversation at the Thanksgiving table might turn toward politics, where your views are well known and not secret. But you might prefer to stay tight-lipped rather than get into a fight. Perhaps you know that you are not going to change your family member's mind on a political issue, so why bother? You might even have the perfect comeback on the tip of your tongue and yet you hold it back. This is very different from, out of fear of being judged, not wanting other people to know who you voted for in a recent election.

You can bite your tongue in conversation for any number of reasons, of which secrecy is only one. Whatever actions you take around keeping your secret, and whatever the context is that surrounds this decision, what is common across all secrets is one thing: you intend for other people to not learn the information. This is what makes something a secret.

THE SECRETS YOU KEEP

When I share my list of the 38 common secrets, people are often interested in how their number of current secrets compares to the average number we see in our research, 13. Often, the question they have in mind is: *Am I more secretive than the average person?* To answer this question, it's important to understand that the number of secrets you have from the list captures not only how many secrets you have, but also *how often you get yourself into the kinds of situations that people tend to keep secret.* So, what makes us more or less likely to land in such situations? And when we do, what makes us more or less likely to keep them secret?

When we start talking about tendencies for secrecy, we bump right up into personality psychology. A common way of measuring personality is to ask about five broad traits: Openness (open to new experiences and to things being complicated), Conscientiousness (organized, disciplined), Extraversion (enthusiastic, social), Agreeableness (polite, eager to please), and Neuroticism (the less polite word for high negative emotion; many prefer to call this "low emotional stability" instead). Just remember the acronym OCEAN if you ever need this information in an emergency.

My research finds that someone who is more secretive (whether having had many experiences from the list or just a few) tends to be less extraverted and less emotionally stable, but more conscientious. The profile of someone more likely to get involved in the kinds of situations that people keep secret, however, is that of someone who is open, extraverted, and emotionally stable, but less agreeable and less conscientious.

This means that while extraversion gets you into more situations that people might keep secret, it is also associated with keeping fewer secrets. Neuroticism and conscientiousness are associated with getting less involved in these situations, but keeping more of them secret.

In general, we see lower well-being among people who are more secretive. But simply having more experiences that people tend to keep secret is not associated with lower well-being.

This is good news. Having many experiences on the list of commonly kept secrets does not mean that you have to suffer. It is keeping these experiences secret that brings potential harm to your well-being, and possibly to your relationships.

You might have noticed that the categories of secrets on our list are those kept by adults. These describe the bulk of our secrets, because people have far more years of adulthood than childhood during which to keep secrets. But adulthood is not where the story of secrecy begins.

CHAPTER 2

The Birth of Secrets

O ne morning, running late for a meeting, I frantically paced around the apartment. I should have already left, but I was still at home, looking at all the surfaces that keys might sit on. This experience was all too familiar, both running late and looking for an object right where I had left it, only to find it mysteriously missing from that spot. Watching me pace around the place, my wife asked me, "Are you looking for your keys?"—as if she had read my mind. In a sense, she had.

People read each other's minds all the time. On a daily basis, we infer what others are thinking and feeling. We do not have direct access to others' minds, but research shows that we are able to understand how they think and feel all the same, from how they act and what they say. We can tell, for example, when someone is excited to join an activity or is joining only begrudgingly. Even with people we don't know well, we can often figure it out. If you heard your coworker say, "Wow, it's really cold in here," you might reasonably infer that this is a request to turn off the AC that you had earlier set to arctic blast. My wife inferred what I was thinking (*where are those keys*) not by clairvoyance, but rather by observing my actions (she had earlier moved the keys to their proper location, of which she came to realize I was unaware).

By the time we enter our adult years, we regularly take other people's perspectives to infer their inner thoughts and feelings. Children can do this too, just not as well, but of course they get better. As they pay more attention to their own mental states, children develop a sharper sense for how to think about *others*' mental states and knowledge, including when that knowledge differs from their own. This, as we'll see, allows them to keep secrets.

HOW UNIVERSAL IS SECRECY?

Recall the lesson of the last chapter: secrecy is not an action but an intention. There are some secrets we never hide in conversation, and a secret can burden us from the moment we intend to hold it back, well before we ever have the opportunity to hide it.

Other animals hide objects, but are humans the only ones who keep secrets? An animal might bury food in the ground for later retrieval. But this is more about keeping their future dinner out of the wrong hands (don't leave food out, or others will take it) than intending to prevent information from entering the wrong minds. To intend to keep a secret, you must be able to understand that something in your head is not necessarily in others' heads.

We previously explored the common secrets people keep; now we'll look at the cognitive abilities that enable us to keep them. From early childhood all the way through adolescence, we will explore the secrets that young people keep, and how their secret keeping changes over time. But before we get to children, we'll first look to chimps.

Like humans, chimpanzees are a social species; their young play, they grin and laugh, they create bonds, they hug and kiss each other. And just as with humans, it's not all fun and games; there are politics within their social ranks, they can become violent, and competing groups have even been observed going to war against one another. Chimps live complicated social lives. But do they also keep secrets from each other? If chimpanzees do keep secrets (or at least, if they try to), this will offer us a glimpse into how other social species hold back from one another, and in turn also reveal which of our experiences with secrecy are uniquely human.

Secrecy in the Wild

To intend to keep a secret, you need to have a mind capable of reasoning about other minds. When animals engage in behaviors that resemble secrecy (such as hiding food), are they ever thinking about other *minds*? For example, if a chimp is just about to retrieve food stashed under some leaves at the very moment another chimp happens to walk by, it will wait a few moments more if the passing chimp is a *dominant alpha male*, who would almost certainly help himself to some of that food. This looks like the chimp is trying to keep the food secret from the alpha chimp. But maybe the chimp is not even considering the alpha chimp's mind. It could be following a more simple rule that it's come to learn from prior experience: don't leave your food exposed to those who would steal it.

To keep a secret from someone, you need to know some piece of information *and* also know that another does not know it. Can a chimp understand when another chimp does not *know* what it knows? To find out, a research team brought two chimps into a chamber. One chimp was just your normal everyday research chimp, but the other was an alpha male to be wary of. Then, some tasty fruit was placed behind either an opaque partition or a transparent partition. When the alpha chimp couldn't see behind the *opaque* partition, the chimp was likely to secretly grab and eat the food when out of the alpha's sight. But when the partition was *transparent*, the chimp was more likely to leave the food where it sat.

So, chimps can understand when something is in another's line of sight and when it is concealed from that line of sight. Does this awareness enable chimps to intentionally conceal from others?

In his book *Chimpanzee Politics*, Emory University primatologist Frans de Waal describes how courting is a bit more direct among chimpanzees than humans. A male chimp might show his interest in a female by showing an erect penis, for example, and De Waal tells the story of a chimp who put on this very display, but then quickly covered it with his hands when a more dominant chimp walked by. Mating chimpanzees have even been observed doing the act extra quietly, like a teen couple with parents just downstairs.

These behaviors look like concealment. But still, it could be that the chimps are *not* reasoning about others' minds.

Perhaps, over time, the chimp has come to learn that quiet sex is safe sex. But chimps will engage in concealment behaviors even in entirely novel lab situations where there is no chance of having previously internalized some relevant rule.

When human researchers act out the role of a dominant chimp—taking food away if a chimp approaches—chimps find ways to sneakily take the food, such as reaching into a chamber that is out of the researcher's line of sight, or opening a door that makes no sound rather than one that makes a metallic clang when opened. It's hard to chalk these behaviors up to rules that chimps have somehow learned from prior experience, as the experimental setups did not look like anything that the chimps had encountered before.

Chimps seem to understand and track what others can and cannot see (or hear), and they can conceal accordingly. But humans do not just hide objects and actions. We also hide information. Can chimps engage in this more sophisticated form of secrecy?

The False Belief Test

To keep *information* secret from another chimp, a chimp needs to understand when the other chimp is unaware of some state of the world. Researchers have created a test for this very situation, called the "false belief test."

The test goes like this: In one condition, a single piece of food is placed in a box for both the normal chimp and the alpha chimp to see. But only the less dominant chimp can see that there is a second piece of food in a second box. So, in this condition, the alpha chimp is only aware of the first piece of food, and so when the trial starts, the alpha chimp should go right to the first box.

In the other condition, both pieces of food are placed in one box—in full view of both chimps—but then, a divider is raised that blocks the alpha chimp's line of sight. Some of the food is then moved from the first box to the second box, but the alpha chimp does *not* see the relocation. And so just like in the other condition, when the trial starts, the alpha chimp should go right to the first box.

In the first condition, chimps *do* understand when the additional food was never in the dominant chimp's line of sight; they know the alpha chimp never saw the second piece of food. And so, the less dominant chimp can grab the food in the second box when the alpha chimp isn't looking.

But, in the second condition, the chimps couldn't seem to grasp the more complicated situation: that the alpha chimp still *falsely believes* that food is only in the first box (not having seen any food get moved to the second box), just as I falsely believed that my keys were still where I left them (because I didn't see them get moved). But unlike my wife, who realized I held this false belief, the chimps did not seem to realize that the alpha chimp held a wrong impression—a false belief—about some state of the world. Even though the chimps could grab the food from the second box when the alpha chimp wasn't looking, the chimps behaved as if the alpha chimp might somehow know that the second box has food in it, and so the chimps didn't dare touch it.

Chimps reason about others' minds in some ways, but not in others. They can understand another chimp's line of sight and where it goes, and so they can act on the knowledge that another chimp does not know about objects unseen. But chimps fail the false belief test. Chimpanzees can't grasp that the knowledge they hold might not be held by others. This places limits on their abilities to engage in secrecy.

CAN BABIES KEEP SECRETS?

At some point, humans surpass chimps' secrecy abilities. We *do* understand that information in our heads is not necessarily in someone else's head. When do children cross this threshold, and when do they start keeping secrets?

Babies may be cute, but they make for lousy conversation partners. We can't just ask them what they are thinking, and they don't yet have the cognitive skills needed to play research games for tasty fruit prizes. So, in their earliest years, to get a sense of what they are thinking about, we have to look to something else; specifically, what they look at. When we and other animals orient our eyes toward something, this means that we are devoting our attention to that thing. When something unexpected happens, for example, both babies and chimps tend to look longer. Based on this idea, researchers have created a false belief test suitable for babies who cannot yet talk or coordinate with others, but who can still look at things.

Imagine a yellow box, a green box, and, in between the two boxes, a children's toy that resembles a slice of watermelon. This was the setup for the study. A woman sits behind the boxes, picks up the toy, plays with it for a moment, and then places it in the *green* box.

Infants fifteen months old watch this scene unfold. Next, a divider is raised that blocks the woman's line of sight, so that she cannot see the boxes—but the infants can still see everything. Then, something magical happens. Thanks to unseen magnets moving below, the toy slides out of the green box and into the *yellow* box!

The divider is then lowered. Next, one group of infants see the woman reach for the toy in *the box in which she last saw it* (the green box where she originally placed it, but where it is no longer). Other infants instead see the woman reach for the toy *where it now resides* (the yellow box), even though she did not see the toy move there.

The infants look longer at the woman when she reaches toward the toy watermelon in its new location, the yellow box. They seem surprised to see the woman act on knowledge that she could not possibly have—as if they expected her to *falsely* believe the toy was still in the green box. But just as we did with the chimps, we should consider the possibility that this has nothing to do with reasoning about others' mental states. Perhaps babies understand that each object has its proper place (toys go in the toy box, books go on the bookshelf), and the woman just showed us that the toy watermelon goes in the green box. Perhaps this is why the infants were surprised to see the woman look for the toy in the yellow box: it's not where it belongs!

We have to look at real behavior, rather than simply where their eyes are pointed, to see if children can truly understand when something is secret. A study with babies aged between sixteen and eighteen months did just this. In one condition, a researcher invites the child to "play a trick" on a second researcher by moving a toy from one box to another box while the second researcher is out of the room, even giggling and making a "shh" sound, with her finger to her lips, to reinforce that this is a sneaky act. In the other condition, there is no trick: the first researcher moves the toy in full view of everyone, and the second researcher never leaves the room.

Next, the second researcher goes to the box *where the toy was originally located*, but struggles to open it. Children at this age like to lend a helping hand if they see an opportunity to do so, and *how* they try to help in this situation is telling.

If the children who witnessed the "trick" understood that the second researcher was not aware the toy had been moved, then, to lend a helping hand, they would *point to the second box where the toy is now located*. But if the toy was moved to the second box in front of everyone, then the adult must be trying to get into the first box for some other reason, and so to help in that situation, the child should try to *help the experimenter push the first box open*. By eighteen months, children behaved in exactly these ways, whereas at sixteen months, they didn't seem to quite get it yet. And so, unlike adult chimpanzees, by eighteen months, infants understood the situation of another person holding a false belief.

Up to a certain age, children behave a lot like their chimpanzee counterparts. Both baby humans and chimps attend to and keep track of another's eye gaze and follow others' basic intentions and goals. And like chimps, toddlers recognize when there are no witnesses.

Where children first surpass chimps is in their ability to grasp that just like objects, *information* can be concealed. This understanding takes some time to develop, however. It emerges gradually and in degrees, over years. And with this progress comes a progression in their use of secrecy.

THE PRE-SCHOOL YEARS

If you're a parent—and even if you're not—you know that pre-school-aged children need to be watched. Once they can walk around, they become able to step away from their guardians' sight. This offers ample opportunity for children to wreak havoc, and so the earliest of kids' secrets often involve them trying to keep their little indiscretions hidden.

Hidden Mischief and Accidents

Children's first efforts to keep secrets are often comical. For instance, a parent told me that his two-year-old loved to blow out candles, even though she knew it was forbidden. One evening, he noticed that a candle was out, and that his daughter was unusually quiet. After he found his daughter hiding, she admitted to blowing out the candle and making a run for it. And a three-year-old helped himself to some Easter chocolate, when he knew he was not supposed to, then hid in a corner of his bedroom, behind the closet door.

In their pre-school years, children will often try to conceal their behavior simply by denying it: one three-year-old denied eating a cookie while having cookie crumbs on his lips, and another refused to admit having gotten into her mother's makeup, with lipstick smeared all over her face.

Unfortunately for parents, children's initial forays into secrecy also typically occur before they become fully toilet-trained. And so, around ages three to five, a very common secret that children try to keep is wetting their pants or their bed (I've also heard stories of kids urinating in other ill-advised places like a bucket or a shopping bag). These incidents are often quickly discovered. But even if a child manages to successfully hide the evidence, a quick line of questioning typically gives the game away immediately, like the three-year-old who successfully hid her accident until she was asked, "But why do you have new pants on?"

Children commonly attempt to conceal their messes at this age. One three-year-old tried to keep a spill secret by sitting on it, not thinking of the wet-butt evidence that would follow. Similarly, a four-year-old tried to hide a drawing he made on the wall by standing in front of it. And after realizing that the bucket of water he had thrown onto his bedroom carpet was nowhere near enough to go swimming, one five-year-old tried to keep his miscalculation secret by simply closing his door.

This stage of secrecy is about on par with that of adult chimps, whose efforts at concealment can be just as comical, although with more adult themes and situations. If an alpha male catches another male chimpanzee mating with a female in his presence (a no-no), the offending chimp might immediately drop its hands over its nether regions as if such an act conceals the entire scene of the crime. Even when the act is successfully done in secret, sometimes the offender can act suspiciously afterward, not unlike the child who refused to answer why she changed into a new pair of pants. In *Good Natured*, De Waal tells the story of a submissive male macaque who successfully mated with a female in secret. Even though the alpha male could not possibly have known about the indiscretion, afterward the macaque displayed unusually submissive behaviors to the alpha male, including even displaying a wide grin.

Unlike chimpanzees, however, children can talk. By age three, children are capable of more complex verbal expression. This means that not only can they issue denials when confronted with evidence of their mischief, but they can also *share* their secrets with others. One three-year-old had a secret crush on a kid at daycare. One day, when his mother asked what he wanted to do for the next couple hours, he looked away as he revealed his secret: he wanted to get married soon.

At this age, children will also share their parents' secrets when they're not supposed to, like the three-year-old who whispered to her father that she had a secret. When he asked what it was, she quietly told him what his Father's Day gift was, having just returned with her mother from the shop in which the gift was purchased.

Crayons and Candles

To understand how secrecy develops throughout childhood, let's return to the false belief test. Young children do show an understanding that others' actions are guided by their intentions, and they seem surprised when others act on knowledge that they could not possibly have. But if you tell a child to imagine that a little boy puts chocolate into a *blue* box, and then, when he's out playing, his mom moves the chocolate to a *green* box, not until age four will children consistently say that upon returning, the boy would expect the chocolate to still be in the first box, the *blue* one.

While young children do show clear hints of understanding others' beliefs, they do not consistently verbalize the correct answer when it comes to false belief tests until around age four. But the issue here isn't poor language skills. It's something else.

In a simplified false belief test, a researcher holds a box of crayons and asks the child, "What do you think is inside?" Children are well acquainted with crayon boxes, and so naturally they say, "Crayons." The researcher then reveals that the box does *not* in fact contain crayons, but candles, and then asks the child, what would *another* person—who has not seen inside the box—think is inside? By ages four to five, children know that the correct answer is crayons: without any information to the contrary, another person would naturally think that crayons are inside a crayon box. At age three, however, children answer that another person would think the crayon box contains *candles*. They can't quite fathom another person not knowing what they do.

Here's where things get really interesting. In another version of the study, after the child registers that there are candles inside the box, the researcher then closes it and asks, "Before, when you first saw the box all closed up like this, what did *you* think was inside?" Three-year-olds will often insist that they *always* knew candles were inside the box.

Perhaps you are thinking that these youngsters are just fooling around and being silly. Surely they can understand that just moments ago they believed the box contained crayons. And yet, that doesn't seem to be the case. In a follow-up study, after the child first guesses the box's contents, but before the candles are revealed, the researcher says, "Look, there's some paper over there. Why don't you get it to draw on with the crayons?" After the child returns with the paper, the researcher opens up the box to reveal that candles are inside the crayon box. As before, the children would often say, incorrectly, that they previously knew that candles were inside the box. When the researchers asked why, then, are you holding a piece of paper, the poor kids had no idea, and typically gave answers like "Because" or "I don't know." They simply could not yet hold in mind that only moments ago they believed that crayons were inside the box, and this is exactly why they retrieved the paper.

What's going on here? To pass the simplified false belief test, children need to understand that the knowledge they currently hold comes from past experience. As children become better able to remember past experiences, they more often pass false belief tests.

It is not that young children are incapable of remembering something from just moments ago, but rather, in order to remember the experience of obtaining new knowledge, they have to pay attention to their inner mental experiences in the first place. Children, at a young age, do not consistently do this.

Toothbrushes and Memories

"I'm going to ask you a question, but I don't want you to say the answer out loud. Keep the answer a secret, okay?" In this study, a researcher asks a five-year-old to silently think about something. "Most people in the world have toothbrushes in their houses. They put their toothbrushes in a special room. Now don't say anything out loud. Keep it a secret. Which room in your house has your toothbrush in it?" The researcher places her finger to her lips and makes a "shh" sound. She lets the child sit silently and after a few moments asks: "What were you thinking about?" Only 31% of the five-year-olds said that they were just thinking about a tooth-brush sitting in a bathroom, and the majority of them, 63%, said that they did not have any thoughts at all. And this oc-curred after they had been taught what it meant to think about something. The children were not paying attention to their own thoughts.

It's not that young children are incapable of attending to their own inner world; they just need some practice. In a final version of the crayon box test, the researcher again asks the child to retrieve a piece of paper for drawing on. For half of the children, the researcher then asks questions like "What will you draw on the paper?" and "What color will you use to draw it?" This helped the children create more elaborate memories of their original false belief. Compared to children who did not spend some time thinking about their inten-tions, children who thought more about what they would draw on the paper more often correctly indicated that they originally believed crayons were inside the box. Without practice, young children do not attend much to their inner thoughts, and this makes tests like these difficult.

The more children attend to their mental processes—paying attention to what they know and what they have learned—the more elaborate memories they form, and the more they recognize that their self exists through time. With these developments, children start to pay attention to which events they were witnesses to, and who was or wasn't there to see it happen. This enables them to understand when they hold knowledge that others do not.

MIDDLE CHILDHOOD

By around age six, children understand that their past experiences accumulate into knowledge and personal memories, which expands their understanding of their self, and their capacity for understanding when something is secret. At the same time, children have a better understanding of other minds, and so they get better at keeping secrets. For example, one six-year-old had been hoarding treats and was hiding the empty wrappers in a small tin that no one would ever look in. She would not have been discovered, except her mother happened to open the tin during a spring cleaning.

At this age, children get better at hiding the evidence, although they are not criminal masterminds just yet. One six-year-old hid dinner food that she didn't like in the floor air vent, which was effective until the family dog began licking the vent. And one six-year-old boy was certainly on the right track in using his "magic" pen to write on the counter with "invisible ink," since it was only visible at a certain angle of sunlight. It could have been the perfect crime, if only he had not written, "Alex is the best."

No child wants to be scolded, and so it makes sense that children will try to use secrecy as a means of engaging in mischief and hiding accidents. But there is much more to secrecy than that (among both children and adults). To have a secret is to have some part of your inner world tucked away from others, whether mischief or merriment. And so, aside from messes and mishaps, I've also heard adorable stories of children's secret accomplishments, ambitions, crushes, and kisses.

With an increased ability for perspective-taking, children think more about their own self, particularly in relation to others. This has the side effect of increasing self-consciousness, and so children start to become embarrassed by some of their proclivities. Imaginary friends, for example, are common, but children can be wary about letting other people know about them. When a parent walks by, children may immediately hush an ongoing conversation with their invisible pals.

As a child grows, so does the potential for having more embarrassing secrets. One parent told me a story about how she noticed that her eight-year-old daughter was quickly tucking her phone away whenever she walked by. This of course was suspicious, and so eventually the mother asked her daughter to hand over the phone. She had been watching videos of people French kissing.

Surpassing Chimps

We've traced the development of children's understanding of minds—others' and their own—alongside their developing use of secrecy. To conceal a physical object or an action, you need to understand what another can and cannot see, or hear. Both children and chimps show this understanding, and will seek to conceal accordingly. But human secrecy departs from chimps' in two ways.

First, unlike the chimp who can't understand that the alpha chimp doesn't know what he knows, children by age four can pass false belief tests in all their forms, and can act on the knowledge that their parents do not currently know about unwitnessed actions. Children develop a sharp sense for how to think about their own mind as well as others' minds, and this enables them to *know* when they have knowledge about some state of the world that others do not. They understand that information which sits in their head is not necessarily in others' heads.

Language helps a good deal here. When parents use more mental processing words with their children, such as "know," "understand," and "remember," this teaches children to attend to their inner worlds, and so children better remember past experiences and more competently understand when they hold knowledge that others do not.

And practice helps a good deal here. Children with more siblings tend to have a sharper sense for how to think about other minds, and going to school provides extensive practice as well. Lessons in reading, writing, and arithmetic focus children on their own mental processes, and the classroom and school playground thrust children into a jungle of peer interactions that force them to pay more attention to others' mental states.

Second, as children gain practice in thinking about others' perspectives, they begin to tell stories. As they better understand others' perspectives, the stories they tell to keep their secrets become more believable; for example, blaming the cat for the broken vase, rather than a ghost. But they do not only tell stories to *conceal*. Children also tell stories to *reveal*. This is the second way in which human secrecy departs from chimps'. We reveal our secrets to trusted others.

Children understand that you can share a secret with someone and still keep it secret from others. One seven-year-old, for example, told her mother that her bracelet was no ordinary bracelet, but her secret lucky charm. But she revealed this secret to her mom on one condition. She made her mother promise to tell no one, not even the family dog.

Sharing Inner Worlds

From the new explosion of peer interactions that children encounter in middle childhood, kids begin to develop friendships: real friendships. Whereas in the pre-school years, friends were simply playmates who happened to be nearby, in older years children can take others' perspectives, and so they notice with which kids they better gel. Friendships become forged in similarity and liking. Children begin to connect more deeply with each other by sharing personal stories.

As children learn that they have an inner world, known only to them unless shared, they begin to section off parts of that world as special, not for keeping all to oneself but for sharing with close others who are deemed worthy. One young child described a secret as "a thing you keep *to* somebody." When asked what a secret is, children will define a secret as something that you can tell your friends and they won't make fun of you. Children will also tell you that what makes someone a *best friend* is that you share your secrets with each other.

And then, things change.

TEENAGE YEARS

"For the most part, my life is totally normal." The movie *Love, Simon* opens with teenage Simon's monologue about how life is good. He is close to his family, he has two best friends he's known since kindergarten, and a new close friend he feels like he's known forever, with whom he does everything. "I have a totally perfectly normal life," he says, "except I have one huge-ass secret." Simon is gay, and he has not come out to anyone.

Simon struggles with his secret. In one instance, he is watching a reality-TV show with his dad, who comments that the star is clearly gay despite the show being about him finding a wife. This makes Simon incredibly uncomfortable. Whenever a conversation reminds Simon of his secret, you can see the discomfort on his face. Yet Simon knows that his friends and family would be accepting and supportive if they were to learn his secret. So why hasn't he revealed it? For one, it can be awkward to talk about romantic attraction. Simon envies straight people who don't have to have *that* conversation; they don't have to come out as straight, as it is just assumed. Also, Simon wonders if his romantic attractions might one day change, and this uncertainty makes him feel even less comfortable with the idea of revealing the secret to anyone.

Eventually, Simon anonymously connects online with someone from his school who is also keeping his gay identity secret. Ironically, this actually makes Simon's secret harder to keep. While he appreciates the emotional support his new contact provides, this makes him think about his secret even more. He has a harder time focusing in school and his mind is often elsewhere, thinking back to his secret, even when spending time with friends.

Who Are You?

By the time children enter adolescence, they have larger social networks, which means more interactions with more people. Teens' growing ability to express themselves and exchange observations, ideas, and stories deepens their relationships with the people around them. This allows them to form meaningful relationships not just with their peers, but also with non-parent adults, such as teachers and friends' parents, when given the opportunity. But what are these teens talking about to all these people? For the most part, themselves.

As children pay more attention to their experiences of the world—and as they experience more of it—their bases of knowledge and personal memories grow. Younger children have plenty of memories, but rather than being organized into volumes and chapters, they are as organized as loose slips of paper thrown into a drawer. Teenagers, in contrast, organize their memories like an autobiography. They develop a narrative that weaves through important past memories, and this provides them with more complex stories to tell. Constructing and sharing this life narrative helps adolescents develop their sense of identity and respond to growing pressures to find and be true to themselves, and this remains true into adulthood.

The teenage years often involve a process of gradually stepping outside of a former self—the one completely enveloped and shaped by family life. In traveling toward these new frontiers of the self, teenagers seek some separation from their family, but unfortunately for parents, this is not always executed with grace. Once they take a bite from the apple of independence, teens will often seek to separate themselves from their parents in every way they can. I've heard parents complain about how, even when on the same outing, teens may put physical distance between their bodies and their parents'—and I remember doing this as well. Behaviors like these are more symbolic than real markers of autonomy, and so of course teens don't stop there. Teenagers, frustrated by having to follow parental rules they don't agree with, quickly realize that there is one big thing they do have control over: what they choose to tell their parents.

Teens can avoid parental control by keeping secrets from their parents about the very behaviors they would seek to control: drinking, dating, ditching school, and so on. And whereas younger children may be willing to discuss secret crushes and kisses with their parents, talk of romance drops sharply in adolescence.

Teenagers draw a line between issues that are under parental control and those that are personal and so deemed as outside of it—though of course parents may seek to weigh in. Issues that teens see as personal tend to revolve around taste and preference, such as clothing, hairstyle, lifestyle choices like diet and bedtime, friendships and relationships, and the one topic that both parents and teens often would rather not discuss with each other: sex.

Discrepancies over what authority parents have in these domains can lead to conflicts, which can be avoided, as many teens come to learn, by holding information back. Up to a point, this is a healthy and normal part of teens' emotional development: a necessary step toward separating themselves from their families, and developing a sense of independence. But when does this become a problem?

It's Complicated

While we have been charting how secrecy develops from its earliest incarnations up through adolescence, we have yet to discuss any potential harms that secrecy might bring.

Here, we are specifically talking about *kids' and teens' secrets around their own behaviors,* and *not* keeping secrets at the behest of others (which usually is a problem, certainly if the request is to cover for someone else's harmful behavior).

Childhood secrets about childhood-sized accidents and indiscretions do not seem to typically take much of a toll, nor is it clear that the occasional lie told to conceal mischief is evidence of a larger pattern of deception.

Things look quite different in later years, however, and perhaps that's because teens can get into more complicated situations and more trouble, such as underage drinking, illegal drug use, and teenage romance. These behaviors can be the backdrop for meaningful stories of friendship and adventure. But if *ongoing problems* are going on unchecked, then keeping these struggles and troubles secret is only going to make matters worse.

Keeping certain secrets from parents—like feeling down, struggling with schoolwork or substance use, or a lingering discomfort, worry, or shame in any domain—can put teens at risk for a host of harms ranging from depression to loneliness to delinquency. We know this from studies of teenage secrecy, but it is difficult to pinpoint whether secrecy itself causes these problems, or whether secrecy is a symptom of another problem. If life is running smoothly, teens will have fewer struggles to keep secret. But for those who have been dealt a tough hand, secrecy may be a reaction to their struggles rather than a cause. Even if secrecy is not where the problems begin, it can compound them.

Of course, what parents know about their teenagers is not just a function of the teenagers' choices. Parents have some responsibility here too. When teens expect that their parents will respond negatively to their disclosures, whether through disapproval, an angry outburst, or some form of punishment, they are more likely to hold back. In a low trust relationship, even well-meaning attempts to encourage sharing can feel like a trap, as teens can construe parents' solicitations for information as attempts to control.

But when teens believe that their parents will express understanding, compassion, and acceptance, and will respond to revelations in a reasonable manner, they are more likely to confide, ask for help, and disclose more freely. As hard as it can be, *in the moment*, for parents to respond in this way (best of luck to you), doing so is the most effective way to keep the lines of communication open.

When teens have healthy relationships with their parents, they are more likely to volunteer stories of their lives: how classes are going, what's happening with their friends, and so on. In healthy relationships, sharing can become habitual. But if parents and their children are less emotionally connected, these conversations may be briefer, and less deep. In unhealthy relationships, *not* sharing can become habitual. Regardless of the quality of their relationships with their parents, however, there will always be things that teenagers let their friends know about, but not their parents.

Relative to parents, who might not understand or who might punish for misdeeds, disclosure to friends is safe. In most cases, friends will be best able to understand the teenager's situation; they may even be going through something similar themselves. And so it is often friends' advice and opinions that teenagers most want, but also their social approval.

Secrecy is a common reaction to feeling uncertain or worried about rejection, but these are exactly the moments when we could most use the support of a trusted other. When it comes to teenage friendships, the holy grail is social validation; teens crave approval from their peers. They act and speak in certain ways and monitor others' responses. When they are feeling down, they can become even more worried about saying the wrong thing.

In the fast-moving social world of teenage life, with ever-changing connections and relationships, fears of rejection and disapproval can eclipse the ability to place trust in others. This is when secrecy is first clearly associated with lower health and well-being.

And this was Simon's struggle. On some level, he so badly wanted to tell his friends and to tell his family his secret. And yet, he wouldn't let himself open up. By keeping hardships and worries secret, teenagers like Simon are closing off the potential for getting the very help and support they need.

This is when secrecy, as adults know it, is born.

CHAPTER 3

Secrets on the Mind

Dale Coventry and Jamie Kunz—two Cook County, Illinois, public defenders—had a secret. They knew that Alton Logan, who had been serving a life sentence since 1982, for murder, was innocent. And they knew this because someone else had confessed to the crime. Despite their anguish over this knowledge, they were legally powerless to share it, and were forced for years to keep the secret locked away while Logan sat in prison for a crime he didn't commit.

Back in January 1982, Edgar Hope and Nadine Smart entered a McDonald's restaurant in Chicago, while Hope's right-hand man, Andrew Wilson, waited in the car. At some point, an argument broke out between Hope and a McDonald's employee. The altercation soon caught the attention of two security guards, who began to approach Hope and Smart. Worried because there was already a warrant out for Hope's arrest on charges of robbery, Wilson, who had seen the altercation unfold through the window, decided to enter the restaurant with a shotgun in hand. Wilson raised the gun, and yelled out to the security guards to back away. And then, in seconds, Hope brought one of the security guards down to the floor, wrested away his gun, pointed the gun to the security guard's head, and pulled the trigger. At the same time, Wilson shot the other security guard in the chest. Then Hope, Wilson, and Smart fled the scene of the crime. The security guard shot in the chest died on the scene, but the other guard survived; he had managed to cover his head with his arm as Hope squeezed the trigger, shielding his head from the bullet.

Nadine Smart later returned to the McDonald's, where police were still documenting the crime. She told them she had witnessed the shooting and wanted to offer information. To protect Wilson, who had committed murder that night—and thus protect herself as an accomplice to the crime—she told the police that Alton Logan, a man she had known since childhood and had instantly recognized, was responsible. She knew that Logan had had his own prior encounters with the police; in 1974, he robbed an elderly man, was caught fleeing the scene in a stolen car, and spent five years in prison, making him a plausible suspect upon whom to pin the crime.

Two days after the McDonald's shooting, Logan, who had been on the straight and narrow for two years following his release from prison, returned home in the evening to learn that police officers had come looking for him. His mother told him to contact the police, which he did right away. They picked him up and took him to the station, where he answered questions through the night and into the early hours of the morning. Then he went home, assuming the whole affair was behind him. But a month later, just before midnight,

Logan heard a knock at his door. He opened it to see multiple guns pointed at his head. He was under arrest.

At the station, the detectives asked if Logan would participate in a lineup. He agreed, and even waived his right to a lawyer: "I don't need a lawyer. Not for doing nothing," he told the officers. Then things went from bad to worse for Logan. The person wounded that day at the McDonald's viewed the eyewitness lineup, and mistakenly pointed to Logan as the shooter.

Wilson, the real shooter, escaped justice that day, but it caught up with him soon after. Only two days after Logan's arrest, two police officers pulled over Andrew Wilson sometime around 2 a.m. An argument broke out, and shots were fired. Both officers died. This time there was no confusion about who was responsible. In the end, Andrew Wilson was sentenced to life without parole for the murder of the two police officers.

Less than a month after the three men were imprisoned, Edgar Hope—the man who shot the McDonald's employee in the arm—told his lawyer that he had never met Alton Logan, and that Andrew Wilson was responsible for shooting and killing the other McDonald's employee. Through Hope's lawyer, this information soon reached Wilson's lawyers, Coventry and Kunz, the two Cook County public defenders. They asked to meet with Wilson, and during that meeting he confessed to the murder. This was the smoking gun that proved Logan's innocence, but Wilson refused to make his confession public.

Without Wilson's cooperation, his attorneys were powerless to remedy the miscarriage of justice that had occurred.

Bound by attorney-client privilege, Coventry and Kunz kept Logan's innocence secret. However, Wilson had given his lawyers permission to reveal his confession after his death. Coventry and Kunz then took the only legal action they could. They wrote and signed a notarized affidavit, detailing that they had learned Logan was innocent and that someone else was responsible for the murder. They sealed the document and put it into a safe, and that's where it sat for the next twenty-six years.

Wilson served his time in prison until the end of 2007, when he died of natural causes, finally allowing Coventry and Kunz to come forward and reveal Logan's innocence, and he was eventually exonerated.

In the end, for nearly twenty-six years, Coventry and Kunz had to keep a secret that they knew could free an innocent man if revealed. The secret weighed heavily upon them. Coventry said he thought about Logan every time he heard of a prisoner wrongly convicted being let free. How could he not? Kunz also frequently thought about the secret; by his own estimate, about 250 times a year. The secret tormented them both, but when asked how, neither told a story of dodging others' questions. When describing the burden they carried, they instead spoke about how frequently the secret returned to their minds.

Whatever actions you take to keep a secret hidden, *to have a secret* is to have an *intention*: keep some piece of information from being known by one or more others. Our minds prioritize anything related to our intentions; this keeps us on the lookout for opportunities to act on those intentions. But this means that our minds will return to thoughts of the secret, even when there is no other person in the room. Like shadows, our secrets can follow us wherever we go. And often, part of the problem is that we are traveling with them alone.

LESSONS FROM COPING WITH TRAUMA

In early November 1983, James Pennebaker, a professor of psychology, then at Southern Methodist University, mailed out surveys to people who had recently lost their spouse. He and his graduate student were granted access to their local coroner's files on recent suicides and accidental deaths. They created a list of surviving spouses, and sent envelopes in the mail to each person on the list. Each envelope held two pages of survey questions and a letter explaining that the researchers were interested in how people cope with trauma. About half of the surveys were filled out and returned.

The researchers asked the participants to what extent they discussed their spouse's death with friends, and the participants also completed a health checklist, indicating whether they had experienced various illnesses and symptoms of poor health, in the period before their loss as well as after. As the researchers anticipated, people reported worse health the year after their spouse died. Irrespective of how they lost their spouse, the participants went from having about one health problem, on average, to having two to three health problems.

Pennebaker's study was not a study on secrets, but a study on coping. In a finding that would alter the course of Pennebaker's career, he found that the *less* participants spoke with their friends about the recent death of their spouse, the *greater* the increase in their health problems. People who *were* talking about their grief seemed healthier than those who were *not* talking about it.

So, talking seems better for the coping process than not talking. But maybe the grievers who discussed their grief simply had more friends to talk to, and that's why they were better off. Pennebaker considered this. Even after comparing people who had the same number of close friends, he found that the participants who spoke less to their friends about their recent loss reported greater increases in health problems.

Pennebaker asked the participants one more question: How much did they find themselves constantly thinking about the death of their spouse? Their responses revealed that the more they discussed their grief with their friends, the less they thought about the death, and the fewer health problems they experienced.

The grieving study leaves us with two puzzles. Why was discussing one's grief associated with less rumination on that grief? And if talking through a difficult experience is better than not talking about it, does this mean that intentionally holding back a secret is harmful?

"Sometimes I hate my job and I don't want to be near sick people."

"Sometimes I feel like I need the recognition that comes with being a hospice volunteer, and I feel guilty."

"I have distanced myself deliberately from some patients and families as a form of self-protection when I've felt emotionally overloaded, even though I felt they needed emotional support themselves."

These quotes come from, as you might have guessed, hospice workers. Dale Larson, a professor of psychology at Santa Clara University, was studying hospice workers around the same time Pennebaker was studying grievers, and he also found that talking about emotional burdens was more helpful to the coping process than not talking. The hospice workers often kept their struggles secret from others, even other hospice workers. They sought to cope with their emotionally exhausting work by not talking about it.

The hospice workers would often blame themselves for feeling worn out, down, and exhausted, and all of this they kept to themselves. Since no one was talking about it, most hospice workers didn't recognize how common their experiences were. Rather than connect over these shared struggles, those who kept to themselves felt isolated and alone. Just like the protagonist in *Love, Simon,* who we met in the last chapter, the hospice workers were *not* opening up exactly when they most needed the support of others. From his studies, Larson became interested in why people become so reluctant to open up about their problems and inner struggles.

"When something bad happens to me, I tend to keep it to myself." "My secrets are too embarrassing to share with others." These are the kinds of statements that Larson asks his participants to agree or disagree with. More than one hundred studies have used his questionnaire, and the conclusions are clear: if your default way of dealing with problems is to keep them entirely to yourself, then you're in for some trouble. Larson finds that in addition to health problems, the habitual use of secrecy accompanies other poor coping strategies, such as trying to avoid problems rather than confront them, and having the expectation that people will respond in the worst ways to disclosures.

By this line of thinking, the grievers we discussed earlier had poor health because they generally didn't have healthy ways of handling problems, and not speaking about their recent loss was one example of their unhealthy style of coping. Larson finds that a tendency to keep quiet about emotional struggles corresponds with feelings of inadequacy, fear of being negatively judged, and being too embarrassed to open up about inner struggles such as these.

A tendency to be too embarrassed to open up, however, is not the same as having a secret. Anita Kelly, professor of psychology at the University of Notre Dame, wondered: What if two people both had the same unhealthy habit of not coming to others for needed help, but one person had a secret at that moment while the other person didn't? Like Larson, Kelly measured the extent to which her participants had an unhealthy habit of holding back rather than dealing with problems, and she also found that a tendency to keep problems hidden was related to poor health. But at the same time, she found that the college students with a major secret said they were enjoying life *more* than the students without a major secret. It really makes you wonder what these college kids were keeping secret. Any guesses? It turned out that the majority of the secrets were about sex and relationships.

These studies suggest that a habit of not coming to others

with emotional struggles is a habit that sits with other harmful ones, like avoiding problems rather than dealing with them. And aside from these unhealthy tendencies, not *all* secrets are harmful. Some can even be exciting, like secrets about sex and romance. By extension, keeping a secret from others is not inherently harmful. So, what separates the secrets that hurt our well-being from those that don't?

This is where I enter the picture. Shortly after starting my job at Columbia, I began working with a graduate student to produce a list of the common categories of secrets people keep: the list I shared with you in the first chapter. Once we had that list, we started providing it to participants, and we asked questions about each secret they had from the list. By looking at a person's whole set of secrets, we can go beyond the question of "Are secrets good or bad?" and instead ask, "*Which* secrets hurt you, and why?"

SECRETS IN THE PARK AND ON THE MIND

For many, the month of September conjures up an image of autumn: students going back to school, light jackets, falling leaves. But in New York City, the summer often rages on. The high density of concrete and asphalt from the buildings, sidewalks, and roads collectively absorb heat throughout the summer, and the temperature rises and rises. On one particular September day, the sun fiercely beat down on a young researcher named Adrien as he arrived at Central Park. Lugging a duffel bag on his shoulder and dragging a cooler behind him, he found a large expanse of grass where people were lounging in groups. His bag was stuffed with paper surveys, clipboards, and pens, and his cooler was filled with ice-cold water bottles.

Adrien had to work up his courage the first time he approached a group of strangers, and a few times after, but he eventually found his groove. He had prepared a small speech explaining that he was a student working on a research project. Would they be willing to complete a short survey? On the sweltering day, Adrien made sure to mention that he had an ice-cold bottle of water to give them in exchange for their participation. As he passed over a clipboard with the survey attached, he hoped the questions wouldn't scare the person away. The first: *Have you ever hurt another person, emotionally or physically, and kept this secret?* The second: *Have you ever used illegal drugs, or were you addicted to, or have you abused, legal drugs like alcohol or painkillers, and kept this secret?* And so on.

We knew that being asked by a stranger in Central Park to complete a survey about your secrets might sound a little odd, if not suspicious. But after learning that our research project was seeking to understand how secrets affect people, so that we could help them better cope with their secrets, most folks were game. Over a few weeks, Adrien approached more than three hundred people lounging in the park who agreed to answer questions about their secrets. Just as we hoped, our participants represented a large diversity of backgrounds and ages, and hailed from diverse places, including tourists visiting NYC from twenty-nine different countries.

Instead of merely counting the number of secrets our

participants kept, we asked specific questions about each secret they had from the list. One such question was precisely the one Jamie Kunz had answered when he estimated that he thought about the secret of Alton Logan's innocence 250 times a year, or an average of five times a week.

When we asked our participants in the park, and several thousand more online, to make a similar estimate, we found that, on average, a current secret returned to mind three times a week. But when we asked people to report on *their most significant secret*—the secret they felt most strongly about—they reported thinking about that secret, on average, twenty times a week: four times as much as Kunz's estimate.

But why even discuss how often we think about a secret? By telling his interviewer how often his mind returned to the secret, Kunz was making the point that the secret weighed on him. And this is what we found when we surveyed people in the park about their secrets. In addition to asking how often they thought about each of their secrets (specifically during moments when they did not have to hide them), we also asked how much each secret hurt their well-being. *The more participants' minds returned to their secret, the more they reported that the secret hurt their well-being.* And this was even true for the secrets they did not need to hide in conversation.

These studies, along with the hill slant studies I told you about in the first chapter, converge on the idea that part of the burden of secrets comes from having those secrets occupy our thoughts. But if thinking about secrets is so harmful, why do our minds so frequently return to them?

THE WANDERING MIND

The human mind has a tendency to wander away from its immediate surroundings. We stare at a work task or chore that must get done, and yet our mind is off somewhere else entirely. Even during a conversation with just one other person, your mind may wander away, which places you in the embarrassing situation of having to pretend that you've been listening, or having to ask for a replay of the last ten seconds. This has nothing to do with how much we like our work or find our friends interesting. Studies estimate that we spend around 40% of our waking hours mind-wandering.

Our mind stays within its lane for whatever it is supposed to be focusing on until all of a sudden it veers off. The thought may not last long, however, as new thoughts constantly march toward center stage and take the spotlight. The reason our minds so often leave the here and now is because we can only lavish attention on something for so long. When you don't busy it with the thing right in front of your eyes, your mind can roam free—such as when looking out the window on your commute, or when taking a stroll on a nice day. Our minds can cover expansive terrains, darting off to faraway places. Is it any wonder that our internal meanderings eventually bump into thoughts of our secrets?

I have a question for you. What were you just thinking about? Really, take a moment and think it through. Your mind has likely strayed away from the words on this page at some point. Perhaps something I mentioned made you think of something else entirely. What was it? If I asked you to write down every thought you had for an entire day, I can guess with great confidence that your list will be filled with thoughts about an errand or a chore, something having to do with money, a mistake you made, your appearance, your health, a political or social issue, your schedule, your career, the weather, food, sleep, and someone close to you. I'm confident that *you* think about most of these things, not because I'm really good at guessing (I'm not), but because I've seen the data. Malia Mason, a professor at Columbia, has spent the past several years buzzing people on their phone throughout the day, asking: Besides the thing right in front of your eyes, what were you last thinking about?

In the seventies, Eric Klinger, a professor of psychology at the University of Minnesota, Morris, estimated that our minds traverse 4,000 different thoughts throughout a typical day. Assuming 16 waking hours, that's 250 thoughts an hour, or 4 thoughts a minute. These numbers come from a study that he conducted after intensely training a small group of research participants to detect when their attention shifted from thought to thought.

In one training exercise, the participants listened to two stories at the same time, one pumped through each ear of a headphone. If you've ever been caught between two conver sations, such as when at a party, you know exactly this expe- rience of strained attention, thinning as it stretches between two targets. The participants were trained to detect when their attention switched between the two stories, and then, at random intervals, the stories would stop and the participants would have to report what they were just thinking about. Finally, participants then carried beepers throughout the day that made a soft ping at random intervals, after which they had to record their last thought and its duration. This training allowed the participants to estimate that, on average—even if flickering back and forth between a small focused few—a thought lasted about ten seconds before the mind pushed another thought onto center stage.

Like hummingbirds, our minds are constantly on the move, but they also tend to visit the same places over and over. The reason you so frequently return to thoughts about your to-do list, money, mistakes, your health, your career, people you care about, and future events is that these relate to your current concerns—that is, your current goals, needs, desires, and plans.

Anything related to your current concerns will easily catch your attention, which is exactly how you want your mind to work. You wouldn't want it to be so focused on something that you wouldn't hear a predator in the bushes, someone calling out your name at a party, or your doorbell. If your mind was never distracted by sounds like these, you might be eaten, fail to realize that your friends are calling you over to their conversation, or never receive your pizza delivery. Your own thoughts usefully distract you in the same way.

Perhaps you have had the experience of standing near a high ledge of some kind—a cliff, a bridge, a balcony—and you thought to yourself, *I could jump,* not because you wanted to, but rather you just imagined this was something that you could do. When researchers asked more than four hundred college students if they ever had imagined jumping from a dangerous height *when near one,* about half of the students answered yes—despite the majority of the students *never* having had a thought about suicide. I've also had the visual image of jumping from a ledge when near one, but why? This turns out to be a consequence of a very sensible anxiety. In a dangerous or anxiety-inducing situation, we imagine the worst-case scenario, the worst possible future, so that we are sure to avoid it. Imagining yourself falling off a ledge is an adaptive thought that you *want* to be distracted by, so that you are extra careful when near a ledge.

When we have an intention, we pay special attention to anything in our environment related to that intention. If a waiter asks whether you want salad or fries, anyone currently with the goal to eat healthy will be reminded of that goal. Of course, just because you remember your goal does not mean you follow through. When I arrive home from a trip and unpack unused exercise clothes from my suitcase, it is impossible to forget the intention I had in bringing them in the first place. We're especially good at remembering unfulfilled goals.

Intentions, current goals, and ongoing concerns are hubs to which the mind wanders for good reason. If you mean to buy a carton of milk, it is actually very helpful when your mind goes *milk, milk, milk* as you enter the grocery store. Likewise, if you have an important deadline to meet, you want to keep that thing top of mind. Just like when you want to be aware of a growling animal in the bushes, or how you want to hear your name at a party, you want anything related to a current goal to immediately draw your attention, whether it is milk beckoning you to buy it, or, with a tinge of sadness, a reminder to ask for salad instead of fries. This is what it means to have an intention; you are on the lookout for ways to uphold the intention.

This feature of human cognition gets to the heart of why our secrets loom so large in our thoughts. Intending to keep a secret means that whenever you are in a conversation, you want to detect anything related to the secret, so that you can be extra careful in what you say. For this reason, you want to be easily reminded of your secrets.

Being on guard and detecting these things early is what helps you play it cool, should a conversation ever get close to your secret. But this increased sensitivity also means that you will be reminded of the secret even when you don't need to hide it in the moment, just as we are frequently reminded of our other ongoing concerns and goals (e.g., a grocery store visit) in moments when they are not directly relevant (e.g., during a work meeting).

Being overly sensitive to something of concern can occasionally lead us to become startled by a shadow behind us, even when it is simply our own. And so, sometimes we find ourselves thinking about our secrets even when we're alone, without a person around to keep the secret from. Having your secret come to mind isn't necessarily a problem, but sometimes when you pick the thought up, it can be difficult to put it down.

STUCK ON A THOUGHT

When we are in a bad mood, our mind doesn't always go to good and helpful places. A natural response to feeling blue is to search for the reason why. What could be bringing you down? Work? Your social life? Your love life? Something else? In your very quest to find the source of your melancholy, you will be hitting play on a montage of sorrow. Negative moods can cause cascading negative thoughts.

Thinking about something over and over can all too easily slide into rumination, which psychologists define not as merely repetitive thought, but persistently *negative* repetitive thought. If we get stuck on a thought, we feel a loss of control, as though we are at the mercy of our own thoughts, and this is why rumination is often paired with feelings of helplessness. To make matters worse, thanks to the fast-moving nature of the mind, we can quickly cycle through many negative thoughts in a very short period of time.

To get a good sense of the toxic nature of rumination, we need to get inside someone's head. The following rapid-fire inner monologue is from the TV show *BoJack Horseman*, which chronicles the life of a washed-up '90s television star. Intermixed among animal puns, zany subplots, and clever storytelling is the heartbreaking tale of BoJack, who is clearly depressed but keeps this secret from almost everyone around him. (I'm not qualified to make clinical diagnoses of people—I'm not that kind of psychologist—but there's nothing stopping me from diagnosing animated anthropomorphic horses.) Here, the mind's inner voice covers substantial self-critical ground in just thirty seconds:

"Piece of shit. Stupid piece of shit. You're a real stupid piece of shit."

"But I know I'm a piece of shit. That at least makes me better than all the pieces of shit who don't know they're pieces of shit. Or is it worse?"

"Breakfast. Oh, I don't deserve breakfast. Shut up! Don't feel sorry for yourself, what does that do? Get breakfast, you stupid fat-ass."

[*eating Oreos at the table*] "Ugh, these are cookies. This is not breakfast. You are eating cookies. Stop it! Stop eating cookies, and go make yourself breakfast. Stop it! Don't eat one more cookie. Put that down! Do not eat that cookie!" [*eats one more cookie*] "I can't believe you ate that cookie!"

This inner monologue begins the episode and it never stops. We hear BoJack's inner thoughts all episode long. It really hurts to be inside his head, but it hurts even more to imagine that this kind of inner monologue is always happening, constantly hidden away, with no ability to pull the plug and make the thoughts stop. Rumination's persistent focus on the negative opens up the floodgates, allowing even unrelated negative thoughts to take hold.

Once your mind wanders toward a secret, it can be difficult to stop thinking of it. The secret can bring to mind negative thoughts, which further push the secret into the spotlight. But what happens when we try to push away thoughts of the secret?

Psychologist Dan Wegner asked participants in one study to do something somewhat strange: Try to not think about a white bear, and please verbalize your thoughts as you have them. Despite their best intentions, the thought still crossed the participants' minds, and they occasionally uttered the to-be-suppressed thought. ("There I go thinking about a white bear again!") And then, in a later phase of the study, when they were allowed to think about a white bear, they thought about the white bear even more. Trying to not think about a white bear *increased* how much participants thought about the white bear. But if we can't even push away thoughts of a white bear, does this mean, then, that we stand no chance against our secrets; that trying to *not* think about them will only make matters worse?

I have good news for you. The act of trying to not think about something doesn't always fail. We are much better at suppressing familiar thoughts than thoughts we've never tried to suppress before. Wegner's participants had never tried to suppress thoughts of a white bear before the study, and so they were bound to fail. They were total amateurs. But when Anita Kelly asked her research participants to identify their most frequently occurring intrusive thought, these participants were later able to suppress it, unlike another group of participants who were asked to not think about a white bear. The more practice you have in trying to tuck away a thought, the better you might be at identifying it, when it comes to mind, and letting it go. Yet if you only try to suppress thoughts that you frequently think about, it may seem as if suppression and rumination always go hand in hand.

The question of whether you can or cannot push away thoughts misses an important point. If you are not talking about a secret with other people, thinking about it is the only way to work through it. In a study of 800 people keeping more than 11,000 secrets, my colleagues and I found that the best predictor of how frequently people's minds wandered to their secrets was not how much they sought to suppress them, but rather how much they *wanted* to think about them. The more important and significant the secret, the more people wanted to spend time thinking about it and figuring out what to do.

While thinking about a secret can be productive, my studies show that people too often simply rehash the details of the secret, or reiterate their regrets. Thinking about a secret is not intrinsically harmful. A persistent focus on the past, however, can be. The more my participants focused on the past when thinking about their secrets, the more mind-wandering to those secrets resembled harmful rumination. But with some degree of forward looking, mind-wandering to the secret was not harmful.

So, what leads people down the dark path of rumination like BoJack's highly negative and self-critical inner monologue? Early life experiences can matter a great deal. When parents rigidly insist on displays of happiness or forbid expressions of discontentment, children may learn to keep their struggles and sorrows to themselves, which in the event of something harmful (problems at home, or worse, abuse) spells real trouble and a need for intervention.

It's best for all parties if parents can make their children feel comfortable opening up, especially in times of need, which as we saw in the last chapter involves responding levelheadedly to admissions: expressing understanding and acceptance. The compassionate response is the helpful one, and the one that keeps the door open for future confessions and requests for help.

If parents fail to provide adequate emotional support for their children, it is likely that those parents are also failing to model effective coping skills. Rather than seek help from others, children may learn to turn inward to deal with their problems. This is why perceiving one's parents as overcontrolling is related to both heightened secrecy and rumination by adolescence. Secrecy allows an escape from parents' criticism, punishment, and anger, but it also precludes the possibility of receiving help when it's most needed.

Even people with relatively happy childhoods and relatively levelheaded parents can slide into unhelpful negative thinking. And they can keep on sliding, and sliding. Rumination contributes to the onset of disorders like depression and anxiety because it acts as a magnifying glass for currently felt negative emotions. When negative self-judgments become global and fail to account for surrounding contexts and details, then self-disgust, self-loathing, and self-pity can sneak in, leaving us feeling inadequate and undeserving. Like the hospice workers that Larson studied, people who tend to keep these emotional struggles to themselves, unaware that their personal difficulties are normal and shared by many others, often develop the habit of turning inward during times of struggle, rather than looking outward, seeking others' help.

I find that the more people tend to turn inward during times of struggle, the more prone they are to ruminate, and both of these unhealthy tendencies are associated with secrets inflicting greater harm as people try to grapple with them. And so, if you feel stuck on a secret, rehashing it and ruminating on how bad you feel about it, this should signal to you that it is time to change course. Rather than look backward into a past that cannot be changed, you can begin to look outward and forward.

FROM INTENT TO BURDEN

The moment you intend to keep something hidden away from someone, or some others, is the moment that something becomes a secret. You might have to hide it in conversation tomorrow, next week, or never, but none of this changes your intent to keep the information unknown to others.

I asked my mother when she and my father first decided to keep secret that my younger brother and I were donor-conceived. Was it shortly after my birth or my brother's? It turned out that neither was the truth. My parents had formed their intention to keep their secret before I was even born. Before there was even a child to keep the secret from, their intention was already solidified.

My parents' secret is a clear example of why we can't define a secret by the actions taken to keep it. Their only intent was to keep the secret from me and my brother. And yet, before it was even possible to conceal the secret, it still had the power to cast doubts, worries, and concerns. What if we looked notably different from my father? What if for some reason my brother and I would one day need to know about our genetic predispositions and the like?

"I started to be more conscious of the secret as you kids got older," my mother told me. "I became more uncomfortable not telling you the truth about your father." Despite this being a secret that required no upkeep or vigilance to keep hidden—no person was asking about the secret—still, her internal discomfort with intending to keep the secret would grow.

We've heard from lawyers keeping a man's innocence secret, from grievers and hospice workers grappling with loss and death, teenagers dealing with insecurities, a government agent discovering secret mass global surveillance, and both a New Jersey mobster and an anthropomorphic talking horse dealing with depression. What this diverse cast of characters have in common is that when they held back from the people around them, they had no difficulty biting their tongues and keeping their stories straight. This was not the source of their pain. Rumination and helplessness, feeling alone and unsupported, worry and uncertainty: these are the common threads in our experiences of secrecy.

Luckily, there are distinct paths forward for better coping. In the next chapter, we'll discuss three.

CHAPTER 4

The Three Dimensions of Secrets

One evening, I arrived at the subway station beneath my apartment to find, to my and my fellow commuters' alarm, a man sitting on the platform with his legs dangling over the ledge, jutting into the path of the train soon to arrive. He waved his arms wildly, making people wary of approaching. He seemed unwell. The scene grew more tense and the onlookers more frantic as the train's arrival time neared, the digital display showing that the next downtown train would arrive in just two minutes. Everyone was frozen, unsure how to help. My wife darted up the stairs to alert the staff at the ticket booth. I started to follow until I could see that many others were trying to do the same. The employee behind the plexiglass, stationed a flight above the scene, didn't seem to have much control over the situation. I ran back down to the platform. My mind raced through terms like "bystander effect" and "Good Samaritan," but those concepts offered me no answers for what to do *now*, how to intervene *in this situation*. It would be so easy to get pulled down, to accidentally fall on the rails, in trying to grab one of his flailing arms. I found myself frozen again, like everyone else on the platform. And then I suddenly wondered, had anyone tried *speaking* to this man?

I didn't have a specific plan, but I thought maybe if I just spoke to him, I could somehow get him off the platform edge. I don't remember what I said exactly, but it was something along the lines of "What's wrong?" He yelped back a series of words, but I couldn't hear where one ended and the next began. From a distance, I reached out my arm and made a beckoning motion with my hand. I said, "Come here. Tell me over here." To my shock, he got up and approached me. He said he needed money for some food, and I gave him some. He walked away, and everyone breathed a sigh of relief, as the train roared into the station.

This experience is not the kind of thing you keep secret. In the days that followed, I shared this story with friends and colleagues. But why? Because it was interesting? I definitely can think of some personal secrets that are interesting, but that doesn't make me want to share them. Because it was memorable? It was a heart-racing experience, leaving me with a vivid and lasting memory, but we have secrets with similar qualities that we don't share with others.

It would be unusual to keep this kind of experience a secret, because most people would agree that my actions were morally good. We are more than happy to share the good in ourselves with others. This is how we become liked, respected, admired.

We judge morality as one of the most important aspects of

a person's character. One of my favorite studies illustrating this was conducted by Nina Strohminger, a professor of legal studies and business ethics at the University of Pennsylvania. She asked participants to imagine a pill that, once ingested, would permanently remove some quality: for example, imagine the pill removed the ability to ride a bike. She asked, from 0% ("same person as before") to 100% ("completely different now"), how much would that change someone? On average, participants said a pill that changed bike-riding ability would only change someone 20%. But what if the pill changed a hardworking person into a lazy person? They said that this would change someone 50%.

Now, what if the pill could turn someone who is a jerk into someone who is not a jerk? What a great pill that would be! People thought that this would change who a person is by 64%. Traits related to morality—those having to do with empathy for others, following the law, doing the right thing—were given the highest percentages, indicating that moral traits were seen as most central to the self. We see morality as fundamental to a person's character, and because we also tend to see ourselves in a positive light, Strohminger finds in her research that most people believe that, at their core, they are moral and good.

If we think of our core selves as moral, and that morality is central to who we are, then what happens when we've done something wrong? I asked Strohminger this very question after learning about her work, and we exchanged a few emails. Based on her findings, we expected that when people did something morally wrong, they would likely see the behavior more as a slip-up than as a reflection of who they truly are; and so they would be more likely to keep that behavior secret, for fear of people drawing the wrong conclusions.

We ran a study where we provided participants with a list of behaviors, and asked them to rate the morality of each behavior. In actuality, the participants were rating the morality of each of the common secrets. How morally wrong is it to hurt someone, to use drugs, to engage in self-harm, to lie to someone, and so on. Next, we showed participants the list again and asked which behaviors they had engaged in, and, for each, how much it reflected who they truly are, deep down. And finally, we asked: Does anyone know about this; is it secret?

The *more* participants thought a behavior was immoral, the *less* they thought it reflected on their character, and accordingly the *more* likely they were to keep it secret. It is reasonable to worry that admitting to morally wrong behavior could make you look bad, which would fundamentally change how a person sees you. But I have some good news for you. While it doesn't always feel this way—such as when you turn on cable news or scroll through internet comments—it turns out that most people believe that, deep down, other people are good.

Even when you do something bad, people are reluctant to suggest this bad behavior is a reflection of who you are, deep down: the real you. Instead of attributing the behavior to your disposition, people who know you well are likely to generate *situational* explanations for why you acted the way you did. If people could get to know you, they would find and see your inner goodness. This isn't some feel-good mantra that I picked up from my time in California, but a relatively universal belief that extends across cultures and to far-reaching corners of the world.

Participants from the United States, Colombia, Singapore, and Russia read stories about different people who seemed to change over time. For example, some participants read about a deadbeat dad, an abusive boss, a "jerk boyfriend." But wait! That was the past, and now the situation is that the dad is caring and involved, the boss is even-tempered, and the boyfriend is now "excellent," respectful *and* affectionate. Other participants read about the same people, but the stories were told in reverse: the caring and involved dad became a deadbeat, the boss became abusive, and the excellent boyfriend became a jerk. In each case, the participants were asked, now that these people are acting differently, to what extent are they being true to the deepest, most essential aspects of their being?

The participants' impressions all showed the same pattern, no matter what country they were from. When the change was in the positive direction, people generally attributed the changed behavior to the deepest and most essential aspects of the person's self. Even the participants who identified as pessimists attributed positive changes to people being true to their core self. However, when the change went the other way, from good behavior to bad behavior, the participants were more reluctant to say that the new behavior reflected the person's true self.

If you feel that you have changed over time, you are likely to see the positive changes as reflective of the real you, and the negative changes as dips in the road: mistakes from which you've recovered. A view of your self as having changed over time, combined with a generally positive self-view, enables the narrative that your former self no longer reflects the latest and greatest version of who you are today.

We care about being seen as moral. We care so much about morality that *how morally wrong we believe a behavior to be* is one of the three primary dimensions of secrets—with the other two being whether the secret *involves our relationships,* and whether the secret relates to our *personal or professional goals.*

By "dimension," I mean a criterion (or criteria) by which you would naturally organize a set of items. For example, if you were going to organize your bookshelf, one dimension by which you might organize books is by *author* name. Another might be the book's *contents* (fiction, nonfiction), or the *color* of the spine (perhaps you are one of *those* people). You might even sort them by multiple criteria (e.g., by color or author name left to right, with fiction books on the top shelves and nonfiction books on the bottom ones, or even on different bookcases).

Now imagine that I asked you to take all your current secrets and organize them on a bookshelf. One dimension your sorting would likely reflect is how immoral you perceive the secret to be; you would put immoral secrets closer together, and far away from the secrets that you don't see as wrong or immoral.

The primary dimensions by which we perceive our secrets hold the key to understanding how our secrets hurt us, and how we should, in turn, cope with those challenges. But before delving into these dimensions in more detail, let's look at how we found these dimensions by which people mentally arrange their secrets. And to do that, we will revisit the New York City subway.

MAPPING THE DIMENSIONS OF SECRETS

Imagine that you are looking at New York City's subway map. Conjure the image as best you can: red, orange, yellow, green, and blue lines cut vertically through Manhattan and dart around the inner green rectangle of Central Park, adjoining and intersecting with gray, purple, and brown horizontal lines that connect Manhattan to the other boroughs. Add a few bends to some of those lines and also a line of lime green, and you have the underground map of the Big Apple.

Now, take away the subway lines and imagine a map of only the stations. On this map, there are no colorful lines: only a series of pins marking the locations of the stations. If you need some help in this exercise of imagination, peek at the following image before reading on.

Even without seeing the hidden underground tunnels that connect the dots, by looking at a map of *just the stations* you can immediately see that most of the Manhattan subway lines run north and south. You can also see that some lines run from the west to the east, connecting Manhattan to the city's other boroughs to the right, where the lines also hook upward in diagonal. By tracing these lines, you could re-create the same basic grid that sits atop the real NYC subway lines. It's simply a matter of connecting the dots.

In our research, we ask participants to do something just like this. In one study, we showed participants the list of

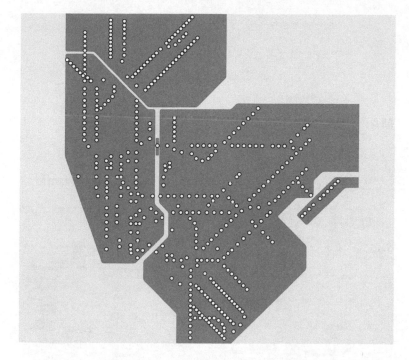

secrets I shared with you in the first chapter, and asked them to arrange the secrets in ways they deemed logical. We then traced out each possible route through the secrets, and asked our participants which of those routes made sense to them—that is, which connected the secrets (or stations) in a logical way. This allowed us to find the orientation of the map and its compass, which produced a map of common secrets.

In the image that follows, you can see that the farther a secret is to the right, the more it has to do with our relationships and social connections. And the closer a secret is to the top of the map, the more the secret relates to our professions and goals. Last, the larger the circle, the more the secret is seen as immoral (a proper 3D map would be *cube*-shaped

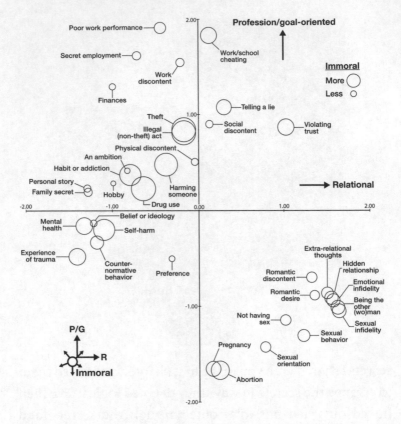

rather than square, but to provide you a sharper image, I've flattened the map to a square here. You can recover the cube shape by imagining that the immoral dimension is represented by depth, such that all the circles are actually the same size, but the secrets that are seen as *not immoral* are far away, and that's why they appear *smaller*).

Certain areas of this map are more associated with harm to your well-being than others, and by knowing where *your* secrets sit within this map, you'll be better equipped to cope with them. So, let's explore this map and the ways in which you can use it to better understand your secrets.

DRAWING THE MAP AND EXPLORING THE THREE DIMENSIONS

Drawing a map is typically straightforward: look at the real world, and spatially locate the landmarks to reflect their relative positions. But how can we obtain the cartographer's vantage point for secrets? To trace out a map of secrets, we need to know where different secrets should be placed relative to one another. For example, should sexual infidelity appear closer to extra-relational thoughts, or to lying to one's spouse about finances? This is where Alex Koch comes into the story. Alex—a professor of behavioral science at the University of Chicago—is an expert in a statistical method called *multidimensional scaling*. That sounds fancy and complicated, but he will be the first to tell you that at its heart, it's quite simple. Because he is German, he will say this in a German accent: all the method does is visualize differences as distances.

If most people agree that one kind of secret—say, a secret hobby—is very *different* from another kind of secret—a secret infidelity—then on the map we should place these secrets far apart. But if a secret—say, a secret preference—is seen as relatively *similar* to another secret—a secret belief—then those secrets should be placed closer to each other. We asked participants to do just that. Our categories of secrets appeared as tiles on a screen. Participants moved the tiles closer to each other if they thought they were similar and placed those they deemed dissimilar farther apart. Then, across all participants, we calculated how far each secret was placed from each other secret, on average. This gave us a table of distances; we knew how far away each "station" was from the other, but not which route to take to get from one to another. This is like knowing all the possible destinations to take a day trip, and the distance between each place, but without a clue as to where the places are or how to get to them.

We next considered every possible way to arrange the secrets in space. This is the kind of task that you hand off to a computer. With the help of an algorithm, we imagined a series of different universes, shuffling the secrets into random spatial positions. We created several 2D spaces, where we scattered the secrets as though on a standard map. But we imagined other worlds, too: 1D spaces that placed all the secrets along a single line, 3D spaces, and even 4D and higher-dimensional spaces that would be impossible to draw on the page.

In each imagined space, the distance between each secret and each other secret was measured by the shortest line connecting the two. By plotting the secrets in a three-dimensional space, we could produce a map that matched the distance table generated by our participants. And so, at this point, we knew the shape of the space we were mapping, and where to place each secret within it.

The final missing piece was a compass; a map is a lot easier to use if we agree on how to align it with the space it is mapping. For example, on a regular map, you want to draw *two* lines at their proper angles (North–South, and East–West), aligning with those directions in the real world. To figure out where we should place our straight lines, we drew "roads" at every possible angle cutting through the 3D space. We then asked our participants to provide a label that explained the order of the secrets as you would pass them along each road.

Before participants provided any labels, the 3D map looked like a jumble of secrets floating in a sphere; we did not know from which angle we should view it. We knew that many *paths* cutting through the space (and the order of secrets passed) would not be logical, just like how, if you placed random lines across the New York City subway map, most would not match the positions of the subway stations. By chance, some random lines *will* pass the stations in the right order, and those were the paths we were trying to find. Looking at the most common labels provided by participants allowed us to eliminate the roads that didn't make sense, and draw only those that participants indicated led to a meaningful sorting of the secrets. This left us with the three primary dimensions.

First, morality: How morally wrong is the secret? The farther you drive in this direction, the more immoral the secrets get. Second, relationships: How much does the secret involve your interactions with other people? The farther you drive in this direction, the more the secrets are about your relationships and social connections. Third, your personal and professional goals: How much does the secret relate to your goals and aspirations? The farther you drive in this direction, the more the secrets are about getting ahead in life.

By examining each possible path through the space of secrets, we exposed our participants to each potential dimension that made up the space of secrets, and participants told us which orientations made the most sense. Like rotating a map in your hands a few times until it looks right, we asked hundreds of participants to look at the map from every angle. In so doing, not only did our research participants create the map of secrets, but they also helped us find its compass.

Secrets perceived as high on the immoral dimension included illegal acts, harming others, drug use, addictions, lying to others, cheating at work, and violating someone's trust. This does not mean that a secret about drug use, for example, is automatically immoral, but rather that people tend to perceive it as such. Similarly, while most hobbies are not typically considered immoral, plenty *could* be considered so, such as big-game trophy hunting. Any category of secret could be perceived with any level of immorality, and those perceptions will differ from person to person; for example, you might have a secret around drug use that you do not perceive as immoral. In mapping out your secrets, these are the kinds of observations you'll be able to make.

Secrets high on the relational dimension often deal with romantic relationships. These include romantic desire, romantic discontent, romantic thoughts about others while in a relationship, and infidelity, whereas secrets low on this dimension are more centered on the individual, including issues of mental health, a personal story, a hobby, and personal beliefs. It turns out that a family secret is seen, on average, as not having much to do with our relationships or how we connect with others. Rather, we tend to place our family secrets in a similar place as other personal stories from our past, and far away from secrets about our social life and our relationships, which are seen as more relational.

The third dimension is about our goals, strivings, and aspirations, which often (but not always) are related to our professions. The secrets high on this dimension include cheating at work, secrets about money and finances, and secrets about sources of income and employment. In most cases, for secrets like these, we can point to an obvious practical reason for having such secrets (to reveal them might hamper the very goals we hope to work toward). The secrets low on this dimension relate to deeply emotional experiences that are much less oriented toward goals and getting ahead in life (e.g., a past experience with trauma, sexual preferences).

When it comes to our secrets, we think about them in three ways: whether they are wrong, whether they involve someone else, and whether they are goal-oriented. A secret can contain any mix of these. Cheating a client or stealing from a business partner, for example, would be immoral, relational, and related to your goals/profession.

Any secret can be placed anywhere on the map. To understand what these dimensions mean for our well-being, we need to look more closely at each.

Morality

They are lying flat on their backs. They're screaming something, cries for help perhaps. Yet under the din of loud machinery, it is hard to make out what they are saying. They are immobilized, their arms tethered to their sides with rope. There are six of them in total, five lined up side-by-side on one set of train tracks, and just a single person on the other train tracks. The train is heading toward the five people. Oh, and I forgot to mention something. You are there too. Your hand is resting on the grip of a large lever. If you pull it, the train will switch tracks, and it will divert away from the five people, but toward the other person on the other tracks. You may recognize this (hopefully) unrealistic scenario as the "trolley problem," a favorite of moral psychologists and philosophers: The train is heading toward five people and will kill them, unless you intervene. All you have to do is pull the lever, and the train will divert, killing only one person. Do you do it?

Dilemmas like these tend to capture the imagination because it is not exactly clear which is the more moral action. Is it wrong to intervene at the cost of someone's life, or is it wrong to let five people die, when the casualties can be limited to just one?

Or what about this scenario: A man enters the supermarket. He breezes past the familiar aisles, not stopping at the displays he would normally browse through if this were his weekly grocery trip. But this is a different trip. He is here to buy one thing only. He purchases a whole uncooked chicken from the deli section, and brings it home. This will be his dinner, but not yet. As always, there is nobody at home to witness him, and he carefully cleans up. He lives alone in his apartment and all the window shades have been shut. He has sexual intercourse with the chicken, thoroughly cooks it, and eats it while watching episodes of *Seinfeld*. Most people would characterize this as deviant sexual behavior (and excellent taste in television), but is it morally wrong? Folks tend to have trouble deciding.

From the late 1950s to the late 1990s, the field of moral psychology was squarely in the domain of developmental psychologists interested in how people develop their sense of morality. For example, a child who once hit other kids and pulled cats' tails could, in later years, become a principled vegetarian and anti-bullying advocate. By the aughts, social psychologists were getting into the game. Psychologist Jonathan Haidt found that people would readily come to answers in moral quandaries like the chicken story, but they had trouble articulating *why* behaviors that seemed harmless or victimless still felt wrong. This phenomenon of "moral dumbfounding," as Haidt called it, inspired a generation of psychologists to more closely examine what makes up our morality.

The trolley problem and the chicken story, however, are far removed from the kinds of moral issues our secrets typically touch on. So what about everyday morality? Psychologist Wilhelm Hofmann (then at the University of Chicago) posed this very question to three other researchers back in 2013, at a meetup for psychologists studying morality in the Chicago area. To find out what morality looks like on a typical day, the researchers texted participants at random intervals, asking them to respond about the past hour. *Did you witness or experience any moral or immoral event, perpetrated by others or yourself?* The researchers found that across a three-day period, participants reported witnessing or experiencing an average of three to four events a day that were seen as either moral or immoral.

Everyday *moral* actions included giving a lost tourist help with directions, giving a homeless person food, and fessing up to a waiter about being undercharged. Everyday *immoral* actions included smoking a cigarette with a child in the car, having a few drinks while on the job, and stealing a coworker's artisanal balsamic vinegar. The researchers also found that we mostly notice our own good deeds and yet pay more attention to others' misdeeds.

When we do notice ourselves committing an immoral act,

one of the most debilitating consequences we can experience is shame: feeling small, worthless, and helpless. We generally believe that immoral acts should be punished, even when the misdeeds are our own. Throughout history, people have engaged in self-punishment as a way to absolve themselves of their sins, and this has even been shown in the psychology laboratory. Brock Bastian, a professor of psychology at the University of Melbourne, conducted a small study that asked one group of participants to recall a time when they rejected or socially excluded someone, and another group to recall a recent everyday social interaction. Next, the participants were moved to a new location supposedly for a different, unrelated study on physical sensation. The participants were then presented with an ice-cold bucket of water, and they were asked to submerge their non-dominant hand and keep it there for as long as they could. If you've ever had to fish a soda or beer out of ice-cold water in a bucket or tub still filled with ice, you know how painful this can be. When participants recalled socially excluding someone, they not only reported their past actions as more immoral (compared to the participants who thought about an everyday social interaction), they also kept their hand in the ice-cold water for longer. But the pain seemed worth it. Those who felt guilty from their memory rated the ice-cold water as more painful, and the brief suffering made them feel less guilty in the moment, as if the small punishment helped rebalance the scales of justice.

When we keep a wrongdoing secret, we escape our justly deserved punishment, and therefore miss out on an opportunity to restore our sense of moral worth. Outside of something like a Catholic confessional, how could you hold yourself accountable for a secret wrongdoing? I posed this question to Brock, and shortly later we designed a study to find out.

We asked one group of participants to think about a secret that they felt bad about and *kept* from their partner, and another group of participants to think about a secret that they felt bad about and had *confessed* to their partner. The still-secret wrongdoings, they said, deserved to be punished, more so than the confessed wrongdoings. And the participants who felt they still deserved to be punished felt less comfortable receiving a kind gesture from their partner or having pleasant experiences like dining with friends, and were even interested in engaging in somewhat painful activities like spending time in isolation, being criticized, and performing a hand-eye-coordination task that could cause temporary wrist pain. The participants behaved as if they were punishing themselves for their secret wrongdoings, paying off their moral debts.

Shame is a particularly painful punishment that people inflict on themselves, prompting feelings of inadequacy, inferiority, and low self-worth. There is unfortunately no magic pill that turns someone from a bad person to a good person, and so, when people see themselves as "bad" or immoral, it may seem as if there is no way to change that. This is why people who feel ashamed can feel powerless and helpless.

In our research we find that the more immoral you judge your own secret to be, the more that secret evokes shame, which is associated with a heightened tendency to repetitively think about the secret, and feel less capable of coping with it. As a result, the more people see their secret as immoral, the more they report that the secret hurts their well-being. Later, we'll explore ways to break free from these toxic feelings of shame.

Relationships

One million affairs. This is the number of affairs *per month* that the dating website Ashley Madison, designed specifically for married individuals, proudly claims to help facilitate. The site's slogan: "Life is short. Have an affair."

Are that many people really having affairs? The current best estimate is that between 20% and 25% of people commit infidelity at some point in their lives, and about 3% of married individuals have engaged in infidelity in a given year (these data come from U.S. surveys of heterosexual married or cohabiting couples, and so the numbers will differ in other countries and contexts). A 2013 Pew survey found that 84% of Americans consider sexual infidelity morally unacceptable. Compare that to the 47% of French individuals, the 64% of Italians and Spaniards, or the 92% of Pakistanis who feel the same way. When it comes to infidelity—as with many other behaviors people tend to keep secret—perceptions of morality vary across country and culture and person.

While sexual infidelity is seen by many as immoral, if not unforgivable, it is also an example of a highly relational secret—both because it involves a romantic or sexual relationship with someone (however fleeting it may be), and also because it is a betrayal of a relationship with someone else.

A few variables predict sexual infidelity. Sexual interest and permissive views on sex, low marital satisfaction, having previously had an affair, and having weak ties to a spouse's friends are all predictors of engaging in infidelity. Infidelity also has a seasonal cycle. During the summertime, infidelity peaks. This could be due to increased travel during these months, creating increased opportunity. People also tend to wear less clothes during the summer, perhaps increasing temptation. There is an increased incidence of infidelity among people who travel for work, and so opportunity for clandestine activity is a clear contributor. While, in the past, psychologists (and everyone else) estimated that men engaged in infidelity more than women, that gap is closing. This could be due to changing societal norms, but a clear contributor is women's increased workforce participation. Going to work yields financial resources and resultant freedom, and also makes it easier to meet people outside of a couple's mutual social network.

Infidelity can mean different things to different people (and implies a closed relationship, as opposed to an open one), but one thing is consistent across cultures and contexts. Across a study of 160 different societies, learning about an infidelity is the number one predictor of divorce, and is an often-cited reason for the dissolution of same-sex relationships (which presumably extends to same-sex marriages as well; the research has yet to be conducted). And so, of course, infidelity is often synonymous with secrecy. While, hypothetically, the sexual aspects of an infidelity may seem like the larger betrayal, a 2002 study found that straight men, straight women, gay men, and gay women reported that the emotional aspects of an actual infidelity hurt them more.

Not all relational secrets, however, are so salacious. Romantic desire while single is a common secret. And relationships in their earliest stages may start off secret, which in the context of something new and exciting can lead to feelings of social closeness and connection. At the same time, if the secrecy hangs around too long, it's associated with reduced relationship satisfaction as well as reduced commitment. And some relational secrets are not at all romantic in nature. For example, violating another's trust, whether a partner, a friend, or a family member, is seen as not just immoral but also highly relational.

Secrets low on the relational dimension are those that center on the self, such as issues of mental health, self-harm, discontent with your physical appearance, a personal habit, a hobby, and a personal belief. These secrets are more individually oriented, and they often make people feel more disconnected from others, more isolated, and alone. Notably, you can have friends and spend plenty of time with others while still feeling isolated, like the hospice workers from the last chapter, who were not opening up to those around them.

A secret can simultaneously offer both high and low social connection. Having a secret affair may foster feelings of increased intimacy and romance with the extra-relational partner, while simultaneously increasing distance from the partner. Such is the power of relational secrets, providing different degrees of social connection and social disconnection.

Profession/Goal-Orientation

Finally we come to the third dimension of profession/goal-orientation, which captures the degree to which a secret is about trying to get ahead in life. At the high end are the secrets related to work, school, and money: cheating at work or school, poor work or school performance, discontent with your profession, secret employment, and anything having to do with your finances.

Financial secrets are often kept from family members, friends, and coworkers, but also spouses. A common secret in marriages is hiding sums of money, or secretly spending money in ways of which a spouse would not approve. Finances are a common source of conflict and strain within a marriage, and keeping financial secrets is one strategy to avoid arguments.

A 2017 TD Bank survey estimated that 36% of couples have monthly arguments about finances, and 13% were keeping a financial secret from their partner, and another research team found that 27% of their participants were keeping a financial secret from their partner. A large phone survey estimated that 6% of individuals cohabiting with their partner have a secret bank account, and 20% reported spending $500 or more without informing their partner on at least one occasion.

It is clear why you might hide a purchase from a spouse, or not reveal the true cost of something: you want to avoid a conflict. Likewise, if you take some shortcut or break the rules at work, it is clear why you would keep that secret. You want to avoid the repercussions. Having a clear instrumental reason for secrecy seems common to the secrets that rank high on the profession/goal-orientation dimension.

Secrets that are very *low* in profession/goal-orientation are the opposite: they don't involve striving, aspiring, or working toward some specific goal. Certainly, people do not aspire to have traumatic experiences or mental health challenges, nor does anyone purposefully seek to be in the position of deciding how to handle an unwanted pregnancy. These secrets are more based in emotion and feeling, whereas secrets high on the profession/goal-orientation dimension tend to be based in logic and deliberation. Many philosophers have considered this contrast. Plato imagined the human mind as a charioteer, who represents reason, sitting at the back of a carriage drawn by two horses, which represent our emotional drives and impulses.

This distinction explains the ordering of secrets along the third dimension. For example, it is difficult to identify the reasons we have the sexual preferences we do. These are not based in logic or reason, but in feeling. Likewise, when it comes to traumatic experiences outside of our control, people understandably do not have an answer ready for "Why me?" While people report having relatively clear thinking about goal-oriented secrets, they often struggle to understand these more emotionally laden secrets. And insight into a secret allows us to better handle the secret and cope with it.

A COPING COMPASS

Our studies find that the more you see your secret as immoral, the more shame you are likely to feel. The more it feels solitary and personal, the more it feels isolating. And the more it is based in emotion rather than logic, the less insight you feel you have into it. By mapping out people's secrets along the three dimensions, we've uncovered the three primary ways in which a secret can hurt you: shame, isolation, and lack of insight. The good news is that these same three experiences point to three different paths forward: three ways in which a secret does not have to hurt you.

The dimensions on our map can be experienced to any degree, and, across each, you can travel in either direction; there are no one-way streets. We can feel more ashamed of a secret or less ashamed. We can feel more alone with our secret or more connected. We can feel uncertain and that we lack insight into our secret or feel that we understand the secret and our reasons for having it.

Knowing where our secrets sit along the three dimensions points us to three distinct avenues for coping. In one study, we gave participants the list of secrets I shared with you in the first chapter. Per each secret that our participants had from the list, we asked questions like "How in control do you feel over this situation?" and "How capable do you feel in your ability to cope with this secret?" Half of the time, before answering these questions, we asked our participants which of the following three options fit their situation best, and all they did was choose one of these three: 1) There is no harm in having this secret (meaning that no one is hurt by the information being contained), 2) This secret protects someone I know (or protects my relationship with that person), and 3) I have good insight into this secret (why I have it and/or how to handle it). To help you pick, consider each option on its own.

IS THERE HARM IN HAVING YOUR SECRET? If you do not feel your secret is wrong or immoral, then you might endorse the statement that there is no harm in having the secret. This is a good thing. And even if you see your previous actions as morally wrong, you may recognize that you can act differently next time; you can learn from your mistakes.

DOES YOUR SECRET PROTECT SOMEONE YOU KNOW? If someone else is in on your secret, perhaps that secret brings you closer together. Or maybe you're keeping the secret to protect someone else, or your relationship with them. In these cases, your secret might be doing some good.

DO YOU UNDERSTAND YOUR REASONS? Even if you acknowledge that you aren't protecting anyone but yourself with your secret, and even if it is causing some degree of harm, simply understanding *why* you are keeping the secret will help you feel more in control of the situation and therefore better equipped to cope with it.

These questions form what I call a "coping compass," because the answers can point you to three distinct paths forward for better coping with your secrets.

COPING STRATEGY 1: Remember that your past mistakes are in your past, and that there's no harm in keeping them there.

I don't like how it feels to bring up this memory from when I was around seven or eight years old. The details are hazy, but I know I was at day camp, and so it must have been summer. I don't remember who first came up with the idea. I want to say it was the other kid, and I was just following along, but I genuinely don't remember if that's true, or simply what I'd prefer to be true. When everyone was off doing whatever activity was on the agenda, he and I snuck into the cubby room and systematically rummaged through people's bags for pocket change, pilfering whatever coins we found, while leaving the paper bills alone. I cannot tell you what possessed us to do this, except that we must have thought we would not get caught. But we did, and in short order. When it was snack time, we purchased a suspicious amount of junk food and frozen treats. Of course it only took a few people noticing that they were missing some change before someone connected the dots. I don't remember the repercussions—it may have been just a scolding. What remains is this memory of this thing that I did.

Unquestionably, our actions were morally wrong, and even at seven or eight years old we *knew* it—and went ahead anyway. Thinking back to this embarrasses me, but an even greater source of discomfort comes from thinking about the straight line that logically connects that past version of me to today-me. If past-me could do such a thing, perhaps this reflects on who I am today, and who I will always be. This is not an idea I like.

But when I look at the coping compass options and ask myself, *Is there any harm in keeping this secret?*, I can't find any harm caused to others at all. Part of the discomfort of thinking about this past failing is that it does not measure up to my personal standards today, but in that same observation is an important insight. I can't hold this past version of myself to my today-self standards, and the secret is totally irrelevant to who I am today.

Moreover, I'm also fairly certain that no one is hurt by not being privy to this information. I certainly would never expect my friends or coworkers to inform me of having done something like this when they were younger. While I see my past behavior as morally wrong, keeping the secret is not.

Thinking about my secret in this way reduces how much it bothers me, and makes it easier to put the thought down and not get stuck on it.

Many of our secrets can fit this mold. Even when you recognize your prior actions as morally wrong, if nobody is being hurt by the information being contained, then keeping the secret is not necessarily morally wrong. Rather than focus on the mistakes of past-you, which cannot be changed, you can focus on the lessons you've learned. You can still feel bad about your past behavior; guilt is a healthy response to judging your *behavior* as wrong. But rather than feel ashamed of your past-self, recognize the improvements you've made and the ways that you've grown. You may have done wrong in the past, but you've changed over time—for the better—and you can continue to do so.

COPING STRATEGY 2: Think of how keeping your secret benefits others.

Ben is not proud of what he's done. He never meant to fall for Mrs. Robinson's daughter, despite her being so obviously more age-appropriate. But it happened anyway, and that's when he realized: Elaine is the one for him. So she absolutely can never know about this. Ben is sure of it: telling Elaine about his prior affair with her mother would bring nothing good, only sorrow and damage to their very new relationship.

You might recognize this as the plot of *The Graduate*. In the movie (and the original book), Ben graduates from college without a clear plan for what comes next, and so he returns to live with his parents in his hometown of Pasadena, California. At his graduation party he encounters a friend of his parents, the seductive Mrs. Robinson, with whom he eventually has an affair. And then her daughter comes home from college, and things get complicated.

Let's step inside Ben's mind. You can imagine yourself however you'd like, but I'm going to imagine myself as a young Dustin Hoffman lying in bed, and despite initial attempts to sabotage it, replaying his unexpectedly delightful date with Elaine:

> Of course, the secret of my affair with Mrs. Robinson hangs over my head. I should have *never* accepted her advances. She is married. She is twice my age. What I did was wrong, I see that so clearly now. Let's be honest, maybe I knew it from the beginning. But what now?
>
> When I look to the coping compass, there is no denying that I've caused some harm with my secret; the sneaking around certainly has not helped Mrs. Robinson's marriage, and if Elaine ever found out, she would be crushed. I don't know why I got myself into this mess, nor do I know how to fix it, but I know that I *am* protecting Elaine, and her whole family, by keeping this secret.
>
> In an ideal world, I would've never had the affair, but there's no putting the toothpaste back into the tube. My best path forward is to continue keeping the secret, and, of course, call off the affair with Mrs. Robinson. Coming clean to Elaine would only serve myself; it might help me clear my conscience, but it will almost certainly cause her

a world of hurt. I can protect both her feelings *and* our relationship by keeping this information to myself. I am confident that keeping the secret does more good than revealing it ever could.

Ben's secret is highly relational. It involves several people. And so revealing it to anyone who knows the Robinsons is risky, and has the potential to backfire and hurt everyone involved: Mrs. Robinson, her husband, Elaine. By keeping the secret, Ben can protect these relationships. Importantly, this calculus might change if Ben believes Mrs. Robinson will reveal the secret, or if he eventually finds himself in a long-term committed relationship with Elaine. He might eventually feel obligated to reveal the secret, as long-term relationships typically come with expectations of honesty and openness, even when the truths are hard. We will revisit this issue in Chapter 6, when we discuss confession.

What impact would revealing your secret today have on the others around you? Could someone get hurt? Even if the journey of your secret ends at confession, think about the good your secret-keeping *currently* brings. You may be protecting not just the people involved, but also their relationships, including their relationships with you.

COPING STRATEGY 3: Recognize that you have your reasons.

For our third coping strategy, we revisit Edward Snowden and his decision to blow the whistle on the NSA's secret mass global surveillance program. Snowden found himself in what seemed like an impossible situation. He believed the NSA program was wrong and immoral, but had signed an oath not to reveal government secrets. He knew that violating that oath could have grave consequences; either he would be thrown in jail, or he would have to leave the country and maybe never come back. But he also understood what was at stake if he stayed silent. In an interview, Snowden said, "I carefully evaluated every single document I disclosed to ensure that each was legitimately in the public interest." Snowden also understood his own values and convictions. In trying to distinguish his actions from those of other leakers, he added, "There are all sorts of documents that would have made a big impact that I didn't turn over, because harming people isn't my goal. Transparency is."

Snowden planned everything to exacting detail in exposing the surveillance program. He took all the necessary precautions in obtaining the evidence, and in planning when and how he would reveal it. Snowden understood the gravity of his decisions, and believed that he was doing the right thing for the right reasons, even if he couldn't always articulate it. "I'd draft manifestos explaining why I'd gone public, but then delete them," he recalls. "And then I'd try writing emails to Lindsay, only to delete them, too. I just couldn't find the words."

While your secrets are likely not as massive as Snowden's, his story offers an avenue for effective coping when the others don't seem available to you. When the decisions surrounding your secret were not easy, there is great comfort in the fact that you have carefully considered all your options, and have made your decision with autonomy, consideration, and care. Understanding the implications of a secret—and why you're keeping it—can bring great clarity. We find that people with insight into their choices and actions feel more capable of coping with their secret.

———

Just like a magnetic compass, the coping compass will not automatically point you to your destination. But you can use it to orient yourself in the right direction. In the vast majority of situations, one of the three strategies can help. We see in our studies that only 1% of the time do people report *very* high shame, isolation, *and* lack of insight. And only 4% of the time do people report *moderately* high shame, isolation, *and* lack of insight. This means that 95% of the time, you have at least one lifeline. The question is: Which is it?

Going through the exercise of deciding which coping resource is most available to you will show you a path forward. When research participants were given the coping compass, they reported increased confidence in their ability to cope with their secrets. And in a follow-up study, we found that the coping compass improved participants' ability to cope, as demonstrated by better daily well-being. So even if you are only taking a small step in the right direction, then you are on your way to better coping.

But we've left an elephant sitting in the room all this while. Although the *burden* of our secrets mostly occurs on our own time, in our own heads, sometimes you *do* have to actively hide a secret during conversations with others, and this brings its own complications.

CHAPTER 5

Concealing Our Secrets

Melody Casson was sixty-seven years old when she finally decided to turn herself in. "I want to tell you something—to tell you everything," she said to a police officer who had come to her doorstep. Her crime had occurred fifty-two years earlier.

On September 6, 1963, when Casson was fifteen, she gave birth to Wayne, her son. Wayne was one of those babies who cried, and cried, and cried. After a bit more than two weeks, she was stretched thin and ready to break. Her mother was in the hospital recovering from a recent operation. Her father was staying home; he was unwell, diagnosed with lung cancer. And her older sister and her sister's fiancé were also home. All were desperate to get some sleep. Yet again, on this night, Wayne would not stop crying, making sleep impossible and fanning the flames of familial tensions. Melody sat next to her infant son on the sofa, and she thought perhaps if she could just muffle his voice for a moment, the sobbing would stop. She took a cushion from the sofa and placed it over his face. She held it there, and Wayne finally stopped crying. When she removed the cushion, it took only a moment's glance to see what she had done. Her son was too still, and his hue was too light. She had inadvertently killed him. She froze, and then yelled to her father as she bolted upstairs.

When the death was reported, Casson claimed she had accidentally rolled onto her eighteen-day-old son when falling asleep, and this is what suffocated him. The coroner's report detailed that Wayne died of asphyxia, caused by "misadventure." Fifty-two years later, she was correcting the record. "I've been living with this all my life," she confessed. At her eventual trial, the judge said, "I accept you've felt guilt every day of your life for the past fifty-two years and I accept in your case there is real, lifelong guilt. You wanted to bring yourself to justice and that is what you have done." The judge continued, "You were fifteen, you were a schoolgirl, you were a child, and you've lived your youth and all your adult life in the shadow of these offences. The punishment inflicted upon you will be lifelong, until the day you die, and nothing I do will alter that." Casson was lucky that she received only a two-year suspended prison sentence (meaning she didn't have to serve the time); had her fate rested in the hands of a less empathic judge, she could have faced jail time.

Why did she come forward in the first place? Why would she risk such significant personal cost just to get something off her chest? "I can't take no more of the pain," she said when explaining her motivation to come forward and reveal the long-hidden secret. But to what pain was she referring? And why would this confession fifty-two years later help ease it?

PROTECTING YOUR SECRETS

Sometimes keeping a secret is merely a matter of staying silent. But other times, silence is not enough; there may be evidence that you need to physically stash away. Infamous Colombian drug lord Pablo Escobar at one point had so much cash on his hands that he resorted to hiding it inside the walls of houses, and in plastic garbage bins buried under the ground. To keep his criminal activities secret, he physically buried the evidence.

While you probably have never found yourself in a situation that required physically burying millions of dollars of drug money, most of us have hidden the evidence of a secret at one time or another, whether burying a candy wrapper in the trash bin, tearing up receipts, or deleting incriminating emails or text messages. In one study, I asked 600 participants to tell me which secrets they had from our list of 38. Across their approximately 7,000 secrets, the ones they most reported needing to physically conceal included secrets around habits and addictions, finances, and possessions.

I find that 26% of secrets require some hiding of the evidence, but the extent to which people physically hide evidence of their secrets is not related to reports of the psychological harm those secrets bring. You might feel a little shady, but the hiding itself doesn't seem to have long-term consequence. After all, hiding at least small objects—say, in the back of a drawer—is not so hard. If you feel the evidence is well hidden, if not destroyed, you can feel more secure that no one will be able to glance upon evidence of your secret.

If you could just put all your secrets in the back of your

sock drawer and forget about them, keeping secrets would be so easy. But it is rare that keeping a secret requires *only* hiding the evidence (just 3% of the secrets in my study). Sometimes, we need to dance around the truth: dodging questions and avoiding entire topics of conversation if they hit too close to home.

By one estimate, on average, we speak around 16,000 words a day. There is the potential for any of your conversations to touch on topics related to your secret, but all you have to do in those situations is bite your tongue. Like the first and second rules of *Fight Club*, just don't talk about it. How sweet it would be (and how much shorter this book) if the secret to managing your secrets was that simple. Of course, it isn't.

When I asked a group of participants about their "complete secrets" (the secrets they have told no one about) and how often they came up in conversations with others, they reported, on average, that someone had asked a question related to their secret about once in the past month (1.2 times a month). Yet those same participants reported that they felt the need to conceal their secret about two to three times, on average, in the past month (2.4 times a month). This means that half the time people feel like they are concealing their secret, no one is actually asking them about it.

I find in my research that the secrets people get most asked about include sexual orientation, their beliefs, their hobbies, and their ambitions. Yet, with the exception of sexual orientation, these aren't the secrets people most frequently conceal, which include: issues around mental health, finances, and romantic discontent. Concealment is not merely dodging others' questions, but just as often it is us stopping ourselves from saying too much.

Rather than a shield to protect from prying eyes, concealment is often a cork we plug in to block our own leaks. So, let's look at what happens when you plug in that cork. Then we'll look at ways to shield yourself from questions you would rather not answer.

Avoiding Conversations

One way to keep a secret is to avoid having any conversations related to it. Sometimes, as was the case for Tony Soprano, this is exceedingly easy. Therapy? He and his associates would never discuss such matters. Their conversations were so far removed from the topic of therapy, it simply would never come up. Outside of someone happening to witness Tony enter a building that houses a therapist's office, nothing could give him away; to keep his therapy secret, all he had to do was never mention it. You might have a secret that's just like this. Perhaps it relates to something that happened a long time ago, or isn't the kind of thing that people often talk about. If no one is going to ask you questions related to your secret, then you don't have to be vigilant for queries and probes. In such cases, all you have to do is keep your mouth shut.

Actively avoiding specific topics of conversation, however, doesn't seem to be a good ingredient for a high-quality relationship. One study asked college students in long-term relationships to what extent they avoided discussing relationship issues, negative behaviors that their partner might disapprove of, and sex. The more the participants avoided these conversations with their partner, the less satisfied they were in their relationship.

Does a relationship in trouble lead to conversation avoidance, or does it go the other way around? One experiment shows the latter is possible: conversation avoidance can be a cause of relationship problems, rather than merely a symptom. When participants were asked to imagine that a friend avoided discussing a topic with them, the participants felt hurt, especially when they imagined the topic had something to do with their relationship. A study of newlyweds also supports the idea that avoiding conversations causes problems. The more the newlyweds believed that their new spouse was concealing from them, the lower their relationship quality was several months later—even when accounting for how much participants said they themselves concealed from their partner. When we think a partner is concealing from us, we feel like we are not included in their lives, and this hurts.

Avoiding difficult conversations can be perceived as a lack of trust in your partner. This doesn't mean that you should discuss every inner thought or disclose every detail about your past. Couples are often quite happy to never discuss the intimate details of prior relationships and prior sexual experiences, for example. "The past should stay in the past" is the idea here, and this explains much of the avoidance around talk of prior sexual experiences. These issues are often deemed not relevant to discuss, and so they feel more like private matters than secrets.

Secrecy emerges when one party not merely fails to mention something, but specifically intends to hold information back. Often, when we keep something from a partner, we believe we're doing so for the good of the relationship—to avoid conflict or spare the other person's feelings. But while keeping a secret from your partner may prevent whatever imagined reaction you expect from them, it has the potential to do more harm than good, especially if your partner senses that you are hiding something. (The nature of this harm is much more obvious when we are the ones who think our partners may be hiding something.)

Suspicions of secrecy, whether right or wrong, are not very good for a relationship. It turns out that how much you trust your partner is based more on your own tendencies to hold back than your partner's. When you believe your partner is secretive, this may license you to engage in kind, creating the potential for a vicious cycle. When one person believes the other is concealing, both parties often lose out: both are likely to perceive the relationship as being of lower quality.

When the road is rocky, we may feel hesitant to open up to a romantic partner, fearing a negative response, or that an admission would only make matters worse. But there is one lever we can use to break these harmful cycles of concealment: trust. Trust your partner, and trust yourself. It may take some courage and vulnerability to initiate a conversation, but don't let your fear close the door on the conversation before it begins. If you think the other person might be unprepared to discuss the issue, avoid blindsiding them. For especially difficult conversations, give some kind of preview or heads-up. And remember, these things take time. You may not resolve everything in a single sitting, but simply starting the conversation is real progress.

Dodging Questions

Maybe you are not ready to talk about something. That's okay. The catch is that we are not the sole arbiters of which topics are introduced into conversation. Another person could always bring up a topic or even ask a question related to your secret. If the topic or question was raised at the wrong time or by the wrong person, what can you do?

If there are multiple people involved in your conversation, count yourself lucky. Two can be intimate, but three (or more) is a crowd. The more people there are in the circle, the less airtime each person gets. Larger conversations don't work very well if multiple people are speaking at once, and so often a few voices dominate, with others barely getting in a word. You can take advantage of this, and choose to be the latter. Just sit back and listen.

But what if a question is directed specifically toward you? The most straightforward way to avoid answering a question is to say that you would rather not speak to the topic. This makes it less likely that the person will press you on the issue—although, reasonably, you might be worried that such a response will be seen as rude or will make things awkward. You don't robotically have to say, "I will not speak to that." There are smoother options. For example, my colleague once told me about a time she was in exactly this position, and her response went something like "Haha, no. You can't ask me that!" By injecting a bit of levity into her response, she gracefully shot down the question without making things awkward.

One especially effective way to deflect a tough question is asking a question of your own. The person's answer can take the conversation in a new direction. A joke can also be very effective. This may seem risky if jokes aren't your thing, but it doesn't even have to be funny to work effectively as a non-answer (I can assure you of this).

Answering a question completely honestly *but as if sarcastically* can also work on occasion, if you can nail the delivery. When Edward Snowden was carrying an old government computer back to his desk with the intention of copying classified files, he was stopped by an IT director who asked him what he needed the old computer for. Snowden replied, "Stealing secrets," and they both laughed.

If you can't think of a jokey response on the fly, an abrupt shift of topic will also deflect attention and keep the conversation moving. You could ask a question ("Hey, what are you up to this weekend?"), or just mention *anything* on your mind that you are happy to chat about ("I need to figure out this thing at work" or "I *need* to get my hands on some food." Alternatively, here's a good one for driving your conversation partner away: "I *just* realized I forgot to brush my teeth!"). Even if you say something offbeat or totally random ("Did you hear that the moon is rusting?"), the person still will have to respond to it, and this can take the conversation far from the initial question.

Questions, jokes, and non sequiturs are effective because each one pushes the original question further into the past. If you've ever wanted to add something to a conversation only to realize you missed your window, then you know how quickly conversations can move from topic to topic. You can use this to your advantage. Once you've pushed the conversation away from the sensitive topic, it will be difficult for others to backtrack.

The most effective way to dodge a question is to do so in a way that feels natural, and responding quickly with whatever comes to mind will go a long way. So don't overthink it. Research suggests that in many cases the other person may not even recognize a question has gone unanswered simply because the conversation has moved past it. It is also the case that most conversation partners are *not* trying to pry into your deepest, darkest secrets, and so even if they recognize that you are dodging a question, they are not likely to press you.

But if for some reason they do, here's one final strategy for your back pocket: thanking them. You may not be especially glad that someone has asked you a question that you wish to avoid, but assuming the interaction is a friendly one, you can be sure that the other person isn't trying to take you down. If you communicate your recognition of the person's well-meaning intentions, declining to answer will go more smoothly. We know this from a study that asked participants to imagine asking a friend about a difficult situation the friend was coping with. Then the researchers asked the participants to imagine that their friend responded in a series of ways.

A response like "Please know that I completely trust you, but let's talk about something else" went over fairly well, because it affirmed the value of the relationship. But even better were responses that expressed gratitude: "It's very thoughtful of you to ask" and "I am grateful for your concern." When the friend thanked them for asking, the participants did not mind when the other half of the sentence was "but I'd rather not get into that" or "but let's talk about something else."

Now of course, if the person asking is your romantic partner, you may not be able to slink away so easily. If you really don't want to speak to the topic at that particular moment, your best bet in this situation is to ask for a postponement of the conversation, and make a plan for when to return to the topic. Your partner should be more receptive to the request if you make it clear that you aren't trying to avoid the conversation entirely; rather, you just want to organize your thoughts first.

Declining to answer a question by saying only that the matter is private, or that it concerns something too personal or embarrassing to discuss, can be insulting to someone who is close to you. We don't like to imagine our friends and loved ones *not* feeling comfortable enough to open up to us. So if you care about your relationship with the question-asker, affirm it in some manner, signaling that *who* is asking is not the problem, but rather, it's the current setting; or say that you need more time. And then, if this buys you time, do use it. Try to organize your thoughts in case you later do decide to talk to someone.

HOLDING YOURSELF BACK

The hard part of keeping many secrets goes beyond just dodging questions or avoiding conversation topics. The truly hard part of keeping many secrets is making sure that *you* don't guide the conversation to the unwanted place.

In his 1963 book *Stigma,* sociologist Erving Goffman wrote about the burdens faced by those with societally stigmatized identities, including those that may not be observable, like having been imprisoned, a religious affiliation, a hidden disability, or one's sexual orientation. While the terms he used to describe these social issues come from another era, the personal struggles of his research subjects sound as familiar as ever. Goffman suggested that we all must decide how much of our true selves to reveal in any given situation or setting. You can choose to express yourself freely or to conceal that which may be judged negatively.

But not everyone is granted the same freedom of expression. The late Katherine Phillips, who was a leading expert on diversity, made this point to me when I first presented my research at Columbia. We were discussing recent company efforts encouraging employees to bring their full selves to work. Her reaction to this idea stays with me today: "Are you kidding me!? I can't bring my whole self to work. I can't speak with my colleagues the way I speak with my friends and family." Some of her colleagues would be surprised to hear that Kathy felt this way; she was well known for her authenticity in her interactions with others, personally and professionally. Yet, as a Black woman—the first Black woman to receive tenure at the Kellogg School of Management *and* the first Black woman to receive tenure at Columbia Business School—Kathy didn't feel that she fit in the traditionally white-male-dominated business school environment. To fit in at work, there was an important part of herself that she felt she could not freely reveal.

As we walk around in the world and traverse our different social spheres—work, home, friends, family—we reveal different parts of ourselves, and certain situations prompt us to hold ourselves back more than others. Drew Jacoby-Senghor, a professor at the University of California, Berkeley, and I have studied the everyday situations that lead people to feel like they can't be their authentic selves. In a sample of over 1,000 individuals, each of whom held at least one marginalized identity, one common situation people experienced was being the only person in the room who looks a certain way, whether along lines of race, gender, socioeconomic status, body shape, beliefs, or something else. Certain questions—like "Where are you from?"—also made people feel like they didn't belong. Even if the person asking was merely curious, the person being asked may interpret it differently: *Why are you here?* Other common situations that threatened participants' identities included being asked to speak on behalf of their social group (e.g., their entire race, gender, or ethnicity), someone assuming something about their upbringing, or someone expressing surprise when they didn't conform to a stereotype.

Our participants pointed to another set of situations where they felt like they couldn't be their authentic selves: everyday conversations. A whopping 81% said they had participated in a conversation that they couldn't bring their full self to, *just in the past week.* We're talking about everyday conversations such as those around entertainment (music, television, movies, books), hobbies, summer travel, and up-bringing. Of course, these conversation topics often make for enjoyable and easy small talk. But if you speak to them freely, you may reveal parts of your self that you feel don't belong.

Across diverse situations and circumstances, we found that when something about the situation made people feel that they did not belong, participants held back parts of their selves, which made them feel inauthentic. Across everyday social interactions, feeling inauthentic, in turn, was associated with daily stress and lower self-reported health.

One especially frustrating aspect of holding back some part of your identity, to avoid appearing different from others (what psychologists call "covering"), is that there is no end in sight. Successfully concealing once doesn't mean you will never have to do so again. Goffman saw this as an added burden for those who must constantly "be alive to the social situation as a scanner of possibilities."

It took fifty years for research to confirm Goffman's prediction: having to monitor social interactions is what makes concealment cumbersome. Clayton Critcher, first at Cornell and later as a professor at Berkeley, asked undergraduate research participants on both coasts to answer a set of interview questions like "Do you ever want to have children?" and "How would you describe your ideal dating partner?" One group of participants was instructed to avoid the words "breakfast" and "therefore," while another group of participants could speak freely. Afterward, the participants performed a task that required some patience as well as some mental rotation (counting how many blocks a shape is made up of, including some hidden from their perspective).

The participants who had to make sure that they did not accidentally let certain words slip performed worse on the follow-up task than did the participants who were permitted to speak freely. Simply having to hold back a few words during conversation had taken a cognitive toll.

Crichter's studies are powerful because they demonstrate how draining it can be to self-monitor, even when avoiding something so trivial as stories of breakfast. But what if you had to hold back something that mattered more? This is what another version of the study sought to find out. Another group of participants was instructed to hide the gender of their ideal dating partner when answering the dating questions. It can be quite difficult to describe someone while carefully stepping over words like *he, she, him, her*. Notably, this also effectively conceals the participant's sexual orientation, and so I have to mention here that these participants were all heterosexual, and as such they did not have much practice hiding their sexual orientation. After the interview, they also later performed less well on the follow-up task, compared to the participants who were allowed to answer the interview questions freely.

When the participants had to hide the gender of their ideal dating partner, not only did they have to watch out for certain words (monitoring for *he, him, she, her*), but they also had to find appropriate replacements (altering course to say *they* or *my partner*). Was monitoring *plus altering* more tiring than monitoring only? In fact, both groups appeared equally fatigued. The additional step of altering language wasn't more tiring than having to pay attention to one's words in the first place.

Of course, performing poorly on a cognitive task given to you as part of an experiment is of little consequence. But there is evidence to suggest that over time, real-life conceal-ment is related to poor health. A study published in the mid-'90s monitored the immune function of HIV+ gay men, every six months for nine years, and found that relative to those who were "out," those who concealed their sexual orienta-tion demonstrated poorer immune function, more rapid progression of AIDS, a greater likelihood of other diseases, and they died sooner. This makes concealment sound deadly. But did those sobering health outcomes result from the men-tally fatiguing effects of concealment, or could they be attrib-utable to something else?

When we look at studies that measure the extent to which people say they conceal parts of themselves, it is often diffi-cult, if not impossible, to disentangle the concealment be-havior from not feeling comfortable enough to be themselves in the first place. To get a sense of where the harm is, we need to know not just whether they concealed, but whether they felt comfortable enough to be themselves. One study cap-tured just this. On each day for two weeks, participants re-cruited from the LGBTQ community in Los Angeles were asked to pay attention to each opportunity they had to dis-close their sexual orientation, and to log whether they dis-closed or concealed it. At the end of the day, participants reported how supported they felt to express their identity. Lastly, two months later, the participants reported on their overall satisfaction with life. The researchers found that concealment was related to lower life satisfaction, but it was also concurrently related to lower social support. And, in looking at people who had the same level of social support, the researchers found that those who concealed fared no worse than the disclosers. Yet when they looked at participants who had concealed to the same degree, having lower social support was associated with lower life satisfaction.

The study suggested that the negative effects of concealment came not from the work people put into hiding their sexual orientation, but rather the negative feelings that go along with not feeling supported enough to disclose it in the first place. And so, as long as you have other sources of support, if concealment protects you from others' unsupportive responses, it may help more than it harms.

CAN PEOPLE TELL?

If you choose to hold back some part of yourself from others, what's the likelihood that any of those others can accurately tell when you are hiding something in conversation? While people *can* tell if you are in a bad mood or if you are looking for your keys, they *can't* tune in to your thoughts.

You have thoughts all the time that you don't share, and people are none the wiser. You had the perfect anecdote to add, but the conversation moved on and you missed your chance. Or you thought of a joke, but then decided not to tell it because it was in poor taste. Or you stopped yourself from making a comment because it wasn't very nice. If people did not make these quick conversational choices, we would live in a less graceful world. While the motivations may differ, the same cognitive skills that allow us to hold back our less than tasteful jokes and our less than useful observations are also what enable us to keep our secrets from slipping out when in conversation with another person.

It may not be technically difficult to hold back a secret in conversation, yet we often worry that our tone, our body language, or our facial expressions will betray us in some way—that others can sense when we're hiding something. But can they? The answer seems to be a qualified no.

In the late 1990s, Laura Smart Richman and Dan Wegner, then at the University of Virginia, conducted a study in which they asked women who reported struggling with an eating disorder to respond to a series of interview questions. The questions started with topics like college life but eventually turned toward self-control, eating habits, and struggles with weight. The participants were instructed to conceal their eating disorder from the interviewer. The study also included women who did not have an eating disorder, who were simply instructed to answer the questions honestly. When research assistants listened to the recordings, they couldn't tell the difference between the participants who were concealing and those who honestly didn't have an eating disorder. They also rated both groups equally on their social skills, engagement, and likeability. Whatever internal struggle was behind the concealment, it stayed beneath the surface.

What if people tried to conceal something that they had no practice at all hiding *and* that required more than stepping over words like "he" or "she"? To find out, Anna Reiman, a professor of psychology, then at the University of Exeter, asked her participants to conceal their college major and to pretend to be medical students, something hopefully none of them had tried to do before. Participants answered some interview questions about college life, and then came a question about their major. One group spoke freely, whereas the other group had to conceal their real major when answering the question. When research assistants watched the videos, and provided their impressions of each participant, what mattered most was how forthcoming, in general, the participants seemed. Whether participants pretended to be a medical student or shared their actual major, the more the raters felt like they got to know the participants from their answers, the more positively they evaluated the participant and the interaction as a whole.

Most experiments on concealment utilize mock interviews as the means of conversation. Yet most real-world conversations don't have an interview format. To test whether concealment would have a negative effect on a less structured conversation, professor of psychology Jin Goh, then at the University of Washington, recruited participants from the university LGBTQ community and asked them to chat about a recent campus issue, concerning university funding of LGBTQ groups on campus. Participants were randomly assigned to conceal or reveal their sexual orientation or gender identity, and they interacted with a participant who identified as straight (who was given no specific instructions about discussing identity). Afterward, they were rated by their conversation partner, as well as by external raters who watched video recordings. According to all perceivers, the students who were concealing looked just as comfortable as those who revealed, and they appeared just as warm and engaged in the interaction.

These studies tell us that most concealments are not detectable—at least not by strangers. Someone who knows you well may be able to detect if something is currently bothering you, or when there is something that you are not saying. But if you are otherwise forthcoming and just being yourself, your interactions will likely go well.

WANTING TO TALK

At the time of her confession, Melody Casson knew that if she had not come forward, there was little chance that anyone would ever find out the truth about the death of her eighteen-day-old son. More than 52 years after the fact, this was never going to come up in conversation on its own, and yet the secret haunted her all the same. Why?

I've never had a secret for anywhere near that long, but it occurred to me that my parents' secret was one they kept for 52 cumulative years (26 each). I asked my mother what it was like keeping her secret for so long. After all, no one was asking about my genetics or my brother's. Like in Casson's situation, it just wouldn't come up in conversation. She said that sometimes she thought about the secret while talking with others and that this felt "awkward," but it was never "conversationally difficult."

But this would change once my brother entered his teenage years and began to ponder which traits he had inherited from his parents and which he developed independently. Mind you, he never suspected that he was not biologically related to our father; he was simply curious about his parents' personality traits, states of health, and so on, and was trying to extrapolate. This is when things "became more problematic," in my mother's words. The more she had the secret on her mind, the more uncomfortable she became not telling us the truth.

There may be times when a secret of yours is so far from your mind, it's as if it doesn't exist. Yet specific events, conversations, or concerns can bring the secret into focus, and when they do, the secret may pull at your attention. As the secret's relevance to your life ebbs and flows, so will its burden. But the burden of a secret comes not only from having thoughts of the secret hang over our heads, but also from having the secret on the tip of our tongues.

It's a mistake to assume that when people keep a secret, this means they don't want to talk about it. When I asked a group of participants to look at our list of the 38 common experiences people typically keep secret, for each experience they had (secret or not), I asked how much they wanted to talk about it with others, and there was actually no difference between the secrets and the non-secrets. Across the experiences that people commonly keep secret, they want to talk about those that are secret just as much as those that are not. What's going on here?

It turns out that two opposing forces are at play. My research participants reported that the more a personal experience felt unresolved, the more they wanted to talk about it. They also reported that, relative to non-secret personal experiences, their secrets felt more unresolved. When it comes to personal experiences that feel unresolved, we seek resolution—often most easily achieved by talking things out with others—but this clashes with the desire to keep the secret hidden. The need for resolution pulls the secret to the top of your mind and to the tip of your tongue, drawing out the disclosure. But your intention—that is, your prior commitment to keeping the information from others—pulls in the opposite direction.

Generally, we want to share with others what is on our mind. When it comes to our secrets, we just don't let ourselves. I find that a useful way to measure this tension is asking people how much they *wish* they would let themselves talk about the secret. Answers to this question correspond with how frequently my participants say they conceal their secrets, even when accounting for how relevant the secret is to their everyday conversations. The more we wish we would let ourselves talk about a secret with others, the more we have to actively stop ourselves from saying too much. Whether you have to dodge questions or not, part of the burden of concealment stems from not letting yourself talk about the things you wish you could.

WHEN DO SECRETS HURT MOST?

As we've seen in the last two chapters, secrets that take up more mental space tend to be more harmful, generating feelings of shame, isolation, and uncertainty. But does this harm come from thinking about the secret, or from concealing it? People's minds tend to wander to the secrets they also frequently conceal in conversation. For example, you might be thinking about a prior moment in which you concealed the secret, or you might be imagining and preparing for a future one. While they sometimes run together hand in hand, concealing a secret during conversations and repeatedly thinking about that secret outside of those conversations are very different experiences.

Concealment can only last as long as the social interaction in which it is taking place, and often only happens for some small fraction of it, during which you will be primarily focused on what you say and how you say it, rather than how bad you feel about the secret.

Outside of social interactions, however, your secret can come to mind at any time. The potential to relive the experience, to think about the secret itself, and to ruminate on how bad you feel is limitless. While thinking about a secret can be productive, it is too often the opposite for one reason: you are doing it alone. When we choose to be alone with something, we often do not find the healthiest ways of thinking about it. We fixate on the negative, blame ourselves, and throw our hands up in the air. Personal tendencies toward unproductive lines of thought will go unchallenged when you keep something hidden away from all others.

Concealment can sometimes be stressful, but not necessarily because of the work it takes to keep the secret hidden. Not feeling comfortable enough to be yourself often hurts more than any conversational gymnastics that a secret requires. Concealment can be uncomfortable in the moment, but then relief settles in and we move on. I find that people tend to construe effective concealment as an accomplishment. You've averted disaster. Seen through that lens, hiding your secrets isn't so bad; from the perspective of your secrecy intention, each concealment is a success.

Melody Casson successfully concealed her secret for fifty-two years, until she couldn't take it anymore and turned herself in. For more than five decades, nobody had even come close to uncovering the secret or asking about it. But her desire to confess remained. The burden of her secret did not stem from worry that someone might somehow find out the truth, but rather that nobody ever would. Her story highlights how concealing a secret is often the easy part. The hard part is having to be alone with the secret. The good news is that you don't have to be.

CHAPTER 6

Confessing and Confiding

One warm summer day in New York's Washington Square Park, among the throngs of college students, chess players, and perspiring tourists, passersby spied something out of the ordinary: a telephone sitting on a wooden table, with a banner reading SECRET TELEPHONE and the suggestion *Get something off your chest.* With the push of a button, you could record your secret, and with another, you could listen to recordings of other secrets previously shared. This curiosity was designed by the artist Matthew Chavez, best known for creating the "Subway Therapy" project.

Over the years, Chavez's installations have periodically appeared in the underground tunnels of the New York City subway, but his biggest project—the one that would make headlines—took place in the days following the 2016 U.S. presidential election, a time when the nation felt more divided than ever. In the 14th Street tunnel between Sixth and Seventh Avenues, Chavez set up a small table that supported piles of pens and pads of Post-its. The idea was simple: write something on a Post-it and add it to the wall. The day after the election, many used the space to express hope. *Let's listen more, Please stay positive,* and *STRONGER TOGETHER We will get through this.*

In the days that followed, the project gained momentum and expanded to Union Square Station, where eventually an estimated 50,000 Post-its covered the walls. The notes ran down corridors and columns and extended from floor to ceiling—from a distance, a dizzying mosaic of neons and pastels, and up close an outpouring of emotion and solidarity.

The project was therapeutic rather than political. Chavez simply wanted to provide people with a place to voice their thoughts and share their emotions. This was also the goal of another project that I happened upon, two years later. On my way to Brooklyn for a concert, I walked through the same subway tunnel where the original sticky-note installation was born to find a banner that read STICKY NOTE SECRETS, and a wall was covered with notes written on black Post-its in silver marker. A black box sat on the table, also marked in silver: *Place secrets on the wall or in the box.* At first, I took a quick photo and then kept walking. I was in a hurry, as most subway passengers are. But then I made a full one-eighty, walked back to the Post-its, and started reading them. And it's a good thing I did, because I ran into Matthew Chavez as he was taking the Post-its down for the night. We set up a time to talk secrets. Three months later the Secret Telephone was born.

The Secret Telephone, in Matthew's vision, is the audio equivalent of his walls of sticky notes; rather than handwritten on Post-its, the secrets were spoken into the receiver. We installed the phone in parks during the daytime. People who walked by were welcome to pick up the receiver, listen to others' secrets, and share their own. We had no idea what to expect when the Secret Telephone was installed for the first time, but people soon lined up to share their secrets. In one recording, someone says that he has been cheating on his partner for the past four years but that he's trying to stop. Another person describes accidentally scraping a piece of artwork. Someone else recalls stealing her best friend's shirt when they were both seven.

People want to confess their secrets. Why wouldn't they? As we've learned, people don't like to be alone with their thoughts, and having a secret can evoke feelings of shame, isolation, and uncertainty. We confide in at least one person for more than half of the secrets we keep, suggesting that, often, the internal suffering isn't worth it.

With good reason, we tend to share our stories with other people. Opening up to others is how we become known, and opening up to others is how *we* learn about ourselves, as well. And then sharing a *secret* with someone, something you wouldn't tell just anyone, can open up whole new worlds of potential: for advice, connection, support—*if* you choose the right person.

STORIES FROM OUR PAST

The past is so special that sometimes you can feel it in the making. We cherish our memories: a moment with friends, an overseas trip, a wedding. More than fifty years after my grandparents' trip around Europe, my grandmother could still tell me story after story about their travels, and recount detail after detail about their experiences. Many of her stories revolved around mishaps, such as getting lost or ordering food to unexpected results. But one memory in particular was special: my grandmother sitting beside my grandfather on a hillside overlooking a lake and the lights of the town below. Next to my grandfather on that hillside, newly married, as dusk settled in, she said to herself, "I will always remember this moment." And she did. She remembered it so clearly that she was able to retell it to me in vivid detail, many decades later. It's wonderful that human memory can work this way, but why do we hold on to all these details?

If you have ever tried to download a movie for later viewing, you may have noticed just how much real estate it occupies on your hard drive. Why do we devote so much mental space to movie-like versions of our personal memories? Why hold on to all the details, the way my grandmother so vividly remembered that moment on that French hillside—the view, the time of day, even what she was thinking about at the time?

Cognitive scientists have long known that our episodic memories—memories of our past experiences (like that moment on the hillside)—are notably different from semantic memories, that is, knowledge and stored facts (for example, that France is the name of a country). Most notably, memories of a past experience are so rich that they include *how* you came to have the memory. This is not the case for semantic memories. You know that France is a country and what makes someone a grandmother, but you don't remember how you came to learn these facts, where you were at the time, and who else was there. Compare this to your memory of your last joyous occasion or celebration. You don't just remember the facts of it, but you can re-experience the event itself. You also know exactly how you came to remember the event: you were there, experiencing it.

The richness of our memory for past experiences is what enables us to share them with others. And we do. A lot. By one estimate, past experiences encompass 40% of what we talk about. Sharing stories of our past helps us learn from each other and learn about each other, which allows us to get close to one another. In *Why We Talk: The Evolutionary Origins of Language,* Jean-Louis Dessalles argues that human communication differs from animal communication in exactly this way: we tell each other stories. Other animals do not describe past experiences to each other, but humans do, and we do so frequently.

Humans are natural storytellers. In the second chapter, we discussed how, as children learn to pay attention to their inner worlds, they get better at recalling past experiences. Storytelling becomes a way to share these experiences with others. As children accumulate memories of past experiences, they increasingly use stories to narrate those events in the order they occurred. Telling stories is how we prove that we saw something to the people who weren't there. Our episodic memory is what gives us the authority to tell stories from our past.

In *Shared Reality: What Makes Us Strong and Tears Us Apart,* Tory Higgins, a professor at Columbia, argues that the purpose of our communication is not only to share knowledge with others, but also to obtain knowledge from them. The world is a complicated place, and we want to see if other people experience it in the same way. I might mention to you that I've been thinking about something in the news, not to give you the latest update on my thoughts, but rather to learn about *your* thoughts on the same matter. We share our thoughts and feelings with other people to hear what they think and have to say. And we want to hear what they think and have to say, so that we can learn if other people experience the world in the same way as we do, to understand how their perspectives line up with our own.

There are an infinite number of topics we can discuss with other people. Yet studies of conversation find that we mostly talk about ourselves. What's this all about? Are humans natural-born narcissists? You can't really be faulted for wanting to talk about your *self*. Your self is, after all, a big part of your everyday experience. You share your self because you want to hear what other people think and have to say. If you keep some part of your self secret from others, then you close the door to finding out what people close to you think; and as we will see, relative to your worst fears, they are likely to have kind things to say.

BEING KNOWN

Humans have a memory system devoted to storing memories so rich that we can share them with others, and much of our communication revolves around telling stories about our past experiences. Sometimes we want to vent, other times we have an important lesson to convey, and on occasion we have questions we plan to ask, but in all cases, sharing our stories with others is how we become known by others.

We may know ourselves better than we know other people, but this doesn't mean we know ourselves entirely. How can we, when the self keeps changing? In one study, I asked participants about their major life events and when they occurred, and I asked how much each contributed to who they are today. The more recent the life event, the more participants said it contributed to who they are. In a follow-up study, I found that whenever their most recent major life event occurred—whether it was moving out of a family home, going to college, starting a new career, moving to a new location, beginning a significant relationship or ending one, a birth or a death, or something else—is about the time my participants say they changed. With time, life changes, and so do we.

There is always more to learn about your self. Conversations with others allow us to listen in, and other people offer us feedback, providing us with new reflections of our selves. If you think you are funny, you might crack a joke among friends. If they laugh, your intuition may be right. If they don't, maybe not; you might want to collect more data. If you lived your entire life alone on an island, besides having no one to compare yourself to, there would be no person to react to you, no person to be your mirror. We need to share ourselves in order to learn about ourselves.

But why do we want to learn about ourselves? With a treasure trove of neuroimaging data, social neuroscientist Diana Tamir, a professor at Princeton, finds that at the most basic level, we want to understand *others* so that we can predict how they will act. When we understand other people's mental states, we know what set of things they are likely to do next, and which are less likely. Learning about ourselves brings the same benefits. Only by understanding yourself well enough to acknowledge that there is some part of yourself you would like to change (something that won't otherwise change on its own) can you work toward that change. Self-knowledge combined with the belief that people can change is associated with optimism, improved decision-making, and effective goal pursuit. There's much to gain in learning about your self, and the best way to do so is to reveal parts of that self to others.

I've just portrayed us all as incredibly self-centered. My apologies. Of course, you do not *only* start up conversations for self-oriented reasons. Most typically, being known by others and knowing them is a two-way street. We want to chat with friends out of genuine desire to keep in touch and stay connected. A long history of research shows that disclosure and social connection go hand in hand.

Disclosure is how we become known, and it is how we connect with other people. This is why across all kinds of relationships—friendships, romantic relationships, family relationships, coworker relationships—the more people mutually disclose, the healthier and happier the relationship.

So, what holds us back?

WHAT WILL PEOPLE THINK?

"I try to put people at ease. I try to stand straight and walk straight. I memorize the relationships between different people and objects in space so that I can refer to them in conversation, and also move fluidly between them. I orient myself towards the people I'm engaging with and gather as much information about them as possible through my other senses so they know they have my attention, so they feel like they're being seen."

This is Sheena Iyengar, professor at Columbia and author of *The Art of Choosing*, speaking to the efforts she puts into her social interactions. Sheena lost her vision at a young age due to a rare congenital disease that causes retinal degeneration. "I try not to make it a thing. But that's not an attitude I've come by easily. In fact, my blindness was very much a thing when I was a child. It was a thing to be ashamed of, to apologize for, to hide in the dark." Up until Sheena was thirteen years old, her family kept her blindness a secret.

"My life seemed to depend on knowing where each person and object was located so that I would not trip, bump, or fumble. And if I did, I had to have excuses and explanations at my fingertips." Why try to keep such a difficult-to-keep secret? *"Log kya kahenge,"* her parents would say. A phrase familiar to those who speak Urdu or Hindi, it translates to: What will people think?

But hiding behind this question has its costs. "My world was already quite small. If I wasn't at home or at school, I was at our *Gurudwara* (temple). Every Friday night, Saturday night, all day Sunday. *'Log kya kahenge'* further circumscribed my activity. And as I lost more and more of my vision, I found myself confined to smaller and smaller circles."

There was one saving grace. Sheena was allowed to reveal

the secret at school. After all, how could she not? The secret was only kept from certain people: namely, other members of her Indian community. "My family members very much wanted to conceal my blindness from other Indians, perhaps because they thought non-Indians were less likely to think of my blindness as a fault or less likely to judge my family for it. Or perhaps they feared far more the censure and rejection of their own community."

But was this fear warranted? Nicholas Epley, a behavioral scientist at the University of Chicago and author of *Mindwise: Why We Misunderstand What Others Think, Believe, Feel, and Want*, finds that when it comes to how other people will react to us, our predictions often miss the mark. He's asked his participants to do things like sing "The Star-Spangled Banner" with a huge wad of gum in their mouths and to sing along as best they can to the rapid-fire lyrics of R.E.M.'s "It's the End of the World as We Know It (and I Feel Fine)." Participants then estimated how well other people would rate their performance, and people were more charitable than they expected them to be. Most folks recognize that those are hard songs to sing, and they adjust accordingly. It is this adjustment we often fail to predict; we forget that other people will take context into consideration.

Once, at a party, a friend of mine said suddenly, "I have something I want to admit to. I didn't do well in college." I asked what prompted her to share this, and after reflexively saying "Oh god, you're researching me," she told me that she suddenly realized her friends wouldn't judge her for it, and just like that, it was a secret easily shed. She had quite accurately recognized that this new information was just a drop in the bucket, one tiny detail out of all our shared experiences and everything else we knew about her.

When you admit something to others, they don't immediately forget everything else about you and only focus on the new information. This is a conclusion that can be hard to come by for the person with the secret. You can get so used to thinking about something in the worst possible way that you forget that there are other ways to think about it. When the only outlet you have to work through a secret is your own mind, you tend not to develop the healthiest ways of thinking about it.

At school, Sheena's classmates knew that she was going blind. "Kids made fun of me, pranked me, tormented me for being so strange. They called me names, put literal obstacles in my path, and either ran away or beat me up when I tried to sit or play with them." School was no haven for Sheena—and yet, she said, she felt freer there: "not from my blindness, but from the burden of hiding it. I felt free to make a life as a blind person, dealing with whatever difficulties that entailed instead of making up an impossible life as a sighted person. And eventually I did find friends and mentors and supporters who taught me to dream of a future in this world, and who helped me to realize many of those dreams in this world."

What will people think? Growing up, Sheena understood that her parents used this phrase as words of caution. "It sounds like a question, but it functions as a command and a warning and an indictment . . . Don't you dare, or how could you, because what will people say." But now, she approaches this question differently. She advises people to "treat it as a genuine question, as a way to compassionately explore what other people *really* will think and say, when you talk with them." Rather than treat this question as a barrier to disclosure, use it as a thought experiment that deserves a thoughtful analysis. And remember that the research findings are clear: People will think more charitably about you than you expect, and this is especially true for the people you are close to and those who know you well. We are quick to judge strangers from one piece of negative information, but this is not the case when it comes to the people who are close to us. In nearly all situations, friends', family members', and romantic partners' impressions of you will not be turned upside down overnight; they won't make global character judgments based on a single piece of information.

When we choose to be alone with a secret, we often focus on the worst and draw the worst conclusions, but people do empathize and people do forgive. It may require courage to bring something up, and if this is the case, know that people will likely recognize it. Revealing something sensitive makes you vulnerable. That might sound terrible, but it conveys trust, and others will recognize when you are placing this trust in them. This is the stuff of intimate relationships, and this is how we get help and support from others.

TO CONFESS OR NOT TO CONFESS

Nikyta Moreno wasn't sure what to make of her husband's strange behavior in the final months of their marriage. "It was like a light switch turned off. He stopped communicating with me and refused to go to therapy. I wondered if he had a medical issue that had changed his personality." Much later, she found out that her now ex-husband had been cheating on her, and all of a sudden the changes to his behavior made sense. Incredibly, it was through his *New York Times* wedding announcement (for his new marriage) that Moreno learned he had been unfaithful. The profile indicated that he had met his new partner in January 2017, when he and Moreno were still married. Keeping his infidelity a secret had made him so evasive and closed off, it was like he was a different person.

If you have a secret like this, should you confess it? This question really combines two different questions: Will the confession provide relief to you, the person confessing? And what effect will your confession have on the other person, and your relationship with them? The answer to the first question is almost always a clear yes. As for the second question: it depends.

It can feel good to unburden yourself of a secret, but what happens after that? Confessing might feel great until the person you confess to breaks down in tears, bursts into anger, or in a dramatic gesture rips off a wedding ring and throws it into the ocean. Of course, only a subset of confessions will end so poorly, but the point here is that the consequences of confessing will largely depend on how the person responds.

People keep secrets to protect their reputations, relationships, and others who could be hurt by the revelation. Yet we feel obligations of openness and honesty toward others, especially with close others. Emma Levine, a professor of behavioral science at the University of Chicago, studies these kinds of dilemmas, and explains that in certain scenarios, there is an obvious norm to follow: kindness. For example, when a friend asks about an outfit in a taxi on the way to the club, there is no point in saying something negative when it is too late to change. Or when someone says, "I miss you," there really is only one nice thing to say in response, which, to be clear, is "I miss you too." (As I once learned the hard way, responding "I don't miss you yet, but I'm sure I will soon!" does not go over well.) Levine finds that when revealing a truth will hurt someone's feelings unnecessarily, people think the right choice is to hold the information back. This is why "white lies" are often seen as more ethical, kind, and compassionate than the unvarnished truth.

The journalist and author A. J. Jacobs once tried to go an entire year without lying (along with following every other rule in the Bible). It didn't go very well. Jacobs tells the story of arriving at a restaurant with his wife to see his wife's friends already dining there. They engaged in some small talk, and his wife's friends suggested that they all get together sometime soon. Jacobs felt that he had to tell the truth, so he said, "You all seem like nice people, but I have absolutely no desire to get together with you . . . I have my own friends I never get to see, so thanks, but no thanks." The response? Jacobs said the friends were visibly offended. And his wife was furious. "So furious she wouldn't even look at me."

When choosing between putting a kind spin on something and being brutally honest, most folks agree that being nice is the preferred option. But this is not the dilemma for many of our secrets. Confessing to having gambled away your child's college fund, having lost your job, or having violated someone's trust is fairly different from giving a generous five-star review instead of your honest three-star one.

In general, we would all be better off with fewer secrets, but at the same time, not every secret should be revealed. So which should you confess?

One factor to consider is whether trust in you could be damaged or destroyed if the other person found out about the secret through some means other than your telling. For example, as DNA testing has become more accessible and affordable, my brother and I could have accidentally found out that we were donor-conceived. "There's No Such Thing as Family Secrets in the Age of 23andMe," reads the title of one article that tells several stories of people unexpectedly learning new information about their family history by using genetic testing services. There's even a whole book on the subject, *The Lost Family: How DNA Testing Is Upending Who We Are*.

Fortunately, my brother and I did not learn about the secret through a genetic testing service. But still, both of my parents later told me that they regretted how we found out. Back in early 2013, while chatting on the phone, my mom began to tell my brother about an argument she recently had, which soon veered into a discussion of how different family members deal with conflict. My mother started telling an illustrative story before remembering the argument in the story was specifically about whether to continue keeping the truth from me and my brother. "And then I realized gosh, you know what, I can't tell him. I was sharing this story, and stopped in the middle because I remembered what the story was about and the context." It is unusual to stop a story midway through, and so my brother asked her to go on. My mother explained that she couldn't, and, when pressed, said it referred to a secret that she had promised to never tell. My brother learned the secret on that phone call, and I found out two days later, late at night after my job interview (luckily, everyone had the good sense to wait until after my interview was over to break the news). But this is not the way my parents wanted us to find out. They imagined telling us at the same time, in person, with all of us together in one room, not slip-ups and late-night phone calls.

So, if it is possible that the secret could come out accidentally or be learned without your telling, then your best bet may be to get out in front of it, and at least have control over the revelation. For the secret that will be found out eventually, or for which concealment is hardly a workable permanent solution, the question of confession should be *when* rather than *if*.

But what about secrets that are less likely to be discovered? A secret from your past that nobody from your present knows about, for example, or some event or act that only you witnessed might fit this category. Why tell anyone these secrets? Perhaps you think the information will challenge someone's incorrect assumption of you, or you want to set some record straight. Perhaps you want to bond with someone over a shared experience, or maybe you just want to unburden yourself.

Aside from what the secret is about, if *keeping* the secret, if holding back itself, will upset the other person—"how could you keep this from me?"—then confessing sooner is better than later. It can be hard to find the right time to have a difficult conversation, but putting it off may only make matters worse. You can build up to it. Let the other person know that you want to talk about something, and maybe even mention what general topic you want to discuss. Even if you don't have the conversation right there and then, this will reduce the shock factor when the conversation happens later.

But what if confessing the secret could damage the relationship; then what should you do? This is the million-dollar question. A classic example of this dilemma is whether or not to confess to an infidelity. Across more than 50,000 research participants, one in three people tell me that they have committed infidelity at some point in their lives. Among the people who have cheated on their partner, about a third eventually confess, about a third never tell a soul, and the remaining third keep it a secret from some people but selectively share it with others.

In his syndicated relationship advice column, Savage Love, writer Dan Savage cautions that while confessing to infidelity could make *you* feel better for having come clean, it could also make your partner feel a whole lot worse. Must you place the burden of this knowledge onto your partner too? If the infidelity happens again, he says, then it is probably a symptom of another problem that needs to be addressed. But if the infidelity was an isolated event and will never happen again, not with that person or with anyone else, Savage suggests that everyone might be better off if only one person bears the burden. If this was a one-time thing, a regrettable mistake, this stone might be better left unturned.

The risk, of course, with Savage's advice is that the secrecy changes your behavior in ways that causes your partner's trust in you to plummet, as was the case with Nikyta Moreno and her husband. No matter what your situation, if you are generally open with your thoughts and feelings, your relationship is likely to be stronger for it.

You should also consider: What would your partner's wishes or preferences be in this matter? I presented the following scenario to three hundred participants in committed relationships. "Imagine that your partner while on a trip (that you were not on) got drunk one night and committed infidelity (had sex with someone). Imagine that this was a huge mess-up, and that it was not a symptom of some larger problem. Imagine it was 100% guaranteed that this was a one-time thing and would never happen again, with this person or with any other person. Would you want to know?" Of the two options, 23% of my participants said, "I would prefer my partner keep this secret from me" and 77% said, "I would prefer my partner confess this to me."

So, should you leave the stone unturned, or should you confess? Would the other person want to know? You don't have to answer these questions on your own. You can confide in someone else, and see what they think.

CONFIDING IN OTHERS

Confiding is like eating your cake and having it too. You get to talk about your secret while still having it remain a secret. Why don't we do this more often?

People rarely ask us about our secrets. On balance, this is a nice feature of the world, but this also means that people rarely *tee up* our disclosures, bringing up exactly the topics we mean to discuss. I've been in this situation before, of wanting to share a secret, waiting for the right time to mention it. And you know what? It never comes. Outside of the situation of mutually sharing secrets (something that only seems to happen during drinking games and after drinking sessions), people do not often open that door for you; you have to open it yourself.

Sheena Iyengar and I wondered what would happen if we opened that door for people. What if we gave them the chance to confide in another—to reveal a secret to someone, specifically someone they were not purposefully keeping the secret from? Our study brought two strangers into the laboratory and had them take turns asking each other getting-to-know-you questions from a list we provided. For example: "If you could take a one-month trip anywhere in the world and money was not a consideration, where would you go and what would you do?"; "Is there something that you've dreamed of doing for a long time? Why haven't you done it?"; and "If you could change anything about the way you were raised, what would it be?"

These questions are a variant of those used in the "Fast Friends Procedure," designed by Arthur Aron and his colleagues back in 1997. The activity leads people to engage in self-disclosure, which, after multiple mutual rounds, creates a feeling of closeness on the spot. Think of the times you've had an enjoyable exchange with a barista, a stranger at a party, or a colleague you don't know well, and felt a real social connection in that moment; that's what this can feel like.

One woman's experience with the Fast Friends Procedure was that she fell in love with her conversation partner, as documented in a wonderful *New York Times* Modern Love essay, "To Fall in Love with Anyone, Do This." Mandy Len Catron explains that "because the level of vulnerability increased gradually, I didn't notice we had entered intimate territory until we were already there, a process that can typically take weeks or months." The research backs up her experience; one study found that when couples asked each other the procedure's thirty-six questions, this increased their feelings of closeness in the moment.

To see whether the Fast Friends Procedure could increase feelings of closeness to the point that people would feel comfortable enough to reveal a personally significant secret, Sheena and I created a faster-paced version of the task, halving the number of questions and dialing up their intimacy, eventually finishing with questions like: "What is your biggest regret?" and "What is your most terrible memory?" So, some pretty heavy stuff.

After the participants asked each other the questions, they were directed to separate rooms, where they were given our list of the 38 categories of secrets and instructed to indicate which they were currently keeping, and which was the most significant one they would be willing to share. Then participants were reunited with their partner and told that they could now chat freely about whatever they wanted—the study or anything else. Oh, and just one more thing: please share your secret with the other person.

After getting to know one another, would participants feel comfortable enough to reveal a personally significant secret? About 50% of the participants shared their secret with the other person. We also asked participants additional questions about their secret, which allowed us to see if the secrets that *were* shared were somehow different from those that were *not* shared.

One complicating factor is that if one person shared, the other person always did the same. So, either both people revealed a secret, or neither did. Possibly this means that feelings of social connection were always mutual, such that some interactions made *both* participants more comfortable opening up. It is also possible that whoever went first determined the outcome for both participants, such that if one person shared a secret, the other person felt obligated to reciprocate and do the same.

Even still, the secrets people talked about were notably different from those they didn't. Before being asked to share the secret, participants were asked what made the secrecy hard. Was it feelings of isolation, inauthenticity, the stress of concealment? Or was it just having to be in your own head, worrying about the secret? It turned out that the more participants were worried about their secret, the more likely they were to share it with the other person. At the end of the study, we also asked participants how it felt to reveal their secret (if they did so). It turned out that the more participants said they were worried about the secret, the better they felt after talking about it.

Now you might be thinking: *But of course, the more the participants were already worried, the more room there was to feel better.* And this is exactly the point. You've likely already found the worst and most harmful way to think about your secret: all on your own. This is the benefit of bringing another person in.

The best time to confide a secret in someone is when you feel comfortable doing so. Maybe you feel comfortable revealing a secret because the person you are talking to just revealed something personal; or maybe the interaction is simply going well, and you see a way in. If you choose the right person, that person will help.

So, who makes for the right person to confide in? I've surveyed thousands of participants on this question. People generally report that someone compassionate makes for an ideal confidant. This makes intuitive sense; someone who is generally understanding and accepting is likely to express kindness and empathy in response to your admission. But I find that people also prefer confidants who are assertive and decisive. The advantage of talking to folks with these qualities is that they will actively help you explore solutions and paths forward, and push you to take them.

If confiding in others somehow entangles them in the problem, or places a difficult burden on them in terms of having to stay quiet, you would be doing those others a favor to *not* confide in them, and instead choose someone else. You could confide in someone removed from it all, even a stranger. I've heard from people who've revealed their secrets to fellow bar patrons, cab drivers, and, of course, therapists. Confiding in a stranger, just as the participants did in my study, can be the best of both worlds. The risks of sharing are minimized— the secret could never get back to anyone you know—but you also get the secret off your chest.

There are advantages to confiding in someone you know, as well. In a study of 200 participants reporting on approximately 3,000 secrets that had been confided in them (across the 38 categories), I found that people frequently reported that they were glad to learn the other person's secret, and felt closer to the person as a result. The vulnerability that comes with opening up—not only does it convey trust, but it's also an act of intimacy. When deciding whom to confide in, remember that people don't typically judge us as harshly as we expect them to, and their impressions of us are nowhere near as fragile as we might imagine. Choose the right person to confide in, and your relationship can likely withstand the revelation.

In addition to choosing someone who will respond in a helpful way, you want to be sure that you confide in someone who will keep your secret safe. Luckily, our research shows that most of the secrets we confide in other people do stay safe with them: 70%, by our best estimate. If that number comes as a pleasant surprise to you, then you're welcome. If that number troubles you, the good news is that you can significantly improve your odds by choosing your confidant carefully.

A natural extension of our human proclivity to exchange stories with one another is that people love to gossip. People bond over gossip: it gives us something to talk about and is a source of entertainment. It also conveys information: it can serve to warn others about bad actors. These two features of gossip, social activity and moral warning, align with the two traits that people do *not* want in their confidants. In our research we see that people are less likely to confide in talkative social butterflies, and they are also less likely to confide in polite goody two-shoes. And so, if a friend of yours is a known blabbermouth, then that friend is probably not your best choice for a confidant. Also, even if someone is generally compassionate and helpful, if you know that the person has a very different set of morals than you, or would be scandalized by your admission, then you might want to look for someone else. In our research, we see that the more someone is morally outraged by your secret behavior, the more likely they are to tell someone else about it. So finding someone with a similar set of morals is a safe bet.

Despite how often we imagine the opposite, confiding usually brings a helpful response. Your confidants have a range of positive responses to choose from: they may offer their ears ("I'm here for you"), validate your experience ("That sucks"), or express sympathy ("Ugh, I'm so sorry to hear that"). Whether the person just hears you out, offers emotional support, or doles out advice, our research shows that having a conversation about your secret is likely to make you feel more confident and capable in coping with it.

FINDING NEW PERSPECTIVES AND CHALLENGING YOUR THINKING

If you have yet to find the right person to confide in, one avenue for better coping with your secret—one that doesn't involve other people at all—is putting your thoughts into written words.

Recall James Pennebaker's study of widows and widowers, which found that talking about grief and trauma seemed to be more helpful than not talking about it. In a follow-up study, he sought to better understand what was helpful about verbalizing one's trauma. Pennebaker wanted to experimentally give people different ways to work through their trauma while eliminating the complication added by how other people respond.

Instead of talking about their trauma with someone, Pennebaker asked undergraduate participants to journal. One group wrote about their *emotions* surrounding a past traumatic event, another group wrote about the *facts* surrounding a past traumatic event, and a third group wrote about both *emotions and facts*. They did this for four consecutive days, and the participants could choose to write about the same trauma across multiple days, or a different one each time. A fourth group wrote about a trivial topic each day, such as describing their home's living room or the shoes they were wearing.

In coordination with the student health center, the researchers were able to obtain a record of how many times the students visited the center due to illness in the months before the writing exercise, and in the months after. Three groups showed an increase in health problems, likely the result of the winter settling in, sending people indoors where it is easier to transmit viruses. But one group showed no increase in health problems over the winter: those who wrote about both emotions and facts surrounding their traumatic experience.

Pennebaker's follow-up studies suggested that acknowledging negative emotions is useful when journaling about one's trauma, but excessively focusing on them is not: since then, the journal just becomes a written record of harmful rumination. Even more beneficial is using words that indicate cognitive processing of the trauma (e.g., *why, how, there-*

fore). Thinking through the causes, reasons, lessons, and insights obtained after a traumatic experience is associated with improved health outcomes.

So, can we skip confiding in someone altogether, and just journal through our problems? I have some bad news for you. Pennebaker is the first to admit that journaling doesn't work for everyone. Journaling, in other words, is not a silver bullet. The ways in which it is helpful are multifaceted and do not always relate to coping with secrets. For example, writing about positive life events seems to have comparable health benefits to writing about negative life events. And writing about coping with a trauma you've never even experienced has comparable health benefits. What is helpful about these writing exercises is that they give people practice in taking multiple perspectives; and they also focus people on things that they might not normally focus on, whether it be the good things in life, different ways to work through stress, or new and productive ways to think about a problem.

If you are not ready to talk to someone, finding a way to take a step back and organize your thoughts is the only way to begin coping. If putting words down on a page sounds appealing to you (or at least not off-putting), you should know that it is more likely to help the coping process if you use the space to step outside of your usual perspective. Try to challenge your usual ways of thinking. And do not overly focus on the negative or the past. Focus more on the present and the future.

There are other ways forward too. You could, for example, reveal your secret by sending an anonymous postcard, as hundreds of thousands of others have done as part of the popular PostSecret project. In 2004, Frank Warren handed out 3,000 postcards outside Metro stops in Washington, D.C., telling people to write a secret on their card and mail it back. By early 2005, he had received enough submissions to regularly post them online, and the secrets—from all over the world—continued to pour in for years. Many of the postcards are moving, some are heartbreaking, and occasionally they are funny (as in the case of the Starbucks employee who confessed: *I give decaf to customers who are rude to me!*). If you don't have a postcard and postage stamp handy, you could also anonymously reveal your secret over the internet (you'll find plenty of websites for this purpose). We find in our research that when framed as an exercise in sharing ("Tell us your secret!"), revealing a secret anonymously over the internet can feel pretty good: the burden of the secret feels lifted in that very moment.

But here's the problem. While releasing your secret into the ether might be easier, and less daunting, it often takes a conversation with another person for you to feel less alone with your secret. Without another *person* in the conversation, it can be difficult to step outside of your usual ways of thinking.

Another person—whether right in front of you, over the phone, or an anonymous internet connection—can give you emotional support that the blank page can't. Another person can challenge your counterproductive ways of thinking, which is especially helpful in working through difficult experiences. And another person can offer practical support and fresh perspectives that are hard to find on your own. This is where journaling—and any kind of revelation that offers no opportunity for another to respond—falls short.

While it may not always feel this way, there *is* a person out there who will listen to you and who will help you. Your confidant may tell you that what you did was wrong or even insist that you need to right your wrongs, but your confidant will also offer you support.

Secrets are not all bad. They can bring people together. In fact, we've been skirting around a really good kind of secret throughout this whole book, and now we can finally turn to it.

CHAPTER 7

Positive Secrets

Take a moment to imagine that you've received amazing news: you've earned an award, a promotion, or something else. Whatever exciting life development comes to mind, imagine something really good has just happened to you. What's the first thing you would do?

In one study, I asked five hundred people this very question, and 76% of them said that the first thing they would do is tell someone the good news. This absolutely eclipsed the second most common answer: 10% said smile or express excitement (cheering, dancing, fist-pumping, yelling "Yes!"). The remaining participants said they would let it sink in and enjoy the moment, express gratitude, or double-check that they heard the news correctly.

Of course, most of us react joyfully to good news, but it is *sharing* the news that is typically top of mind. I once witnessed my colleague storm off in a rage, yelling expletives, upon learning that his paper had been accepted to be published in the journal *Science* (the equivalent of getting your paper admitted to Harvard). It turned out that his frustration stemmed from the fact that his schedule was packed all day, and so he could not *at that very moment* call his partner and plan their celebration.

It doesn't have to be the most amazing news of all time for you to want to share it. In a follow-up study, I asked my participants about all the *good things* going on in their lives, and they intended to tell others 96% of the approximately 3,000 pieces of good news they collectively reported. While this news included major accomplishments and exciting life developments, it also included "small wins" like completing a task, receiving a recognition of some kind, starting a new project, acquiring a new possession, coming up with an idea, finding a lost object, or learning of a positive event on the horizon. These are the kinds of things we want to tell people about.

With each retelling, we get another chance to celebrate the good news. When we share our joys with others, we keep the good times rolling. But we don't always share every piece of good news right away—sometimes we keep something under wraps, waiting for the perfect moment to make our reveal.

This feels like the right time to admit that I've kept a secret from you. Back when I shared the map of the commonly held secrets, I didn't mention that the map was missing two secrets (from our list of 38). Those two secrets—marriage proposals and other planned surprises—weren't included in the map because they sit in a different space entirely, a category unto itself.

When it comes to many "positive secrets," the whole point of the concealment is some big future reveal, and looking forward to that revelation has its own benefits.

BREAKING THROUGH THE EVERYDAY AND LOOKING FORWARD TO THE FUTURE

We don't really look so much to the future when it comes to our secrets. Even when we build cunning defenses and clever cover stories to keep some experience hidden, these exercises are more backward-looking than forward-looking, as they push attention onto the past events (and perhaps alternate versions of them). Not until we begin considering the act of revealing a secret do we typically start thinking about its future.

When it comes to *positive* secrets, however, we are much more likely to look ahead to that moment when our secret will be revealed. Thinking about revealing something positive can make that positive shine more brightly, not only during the lead-up to the reveal, but often afterward too.

Something Good on the Horizon

Imagine you have plans to travel to a nearby city for two weekends in a row. One weekend you are visiting a somewhat abrasive family member to whom you are not very close. You do not expect to enjoy yourself on the visit, but it's just one of those things that you have to do. The other weekend, you will get together with good friends that you haven't seen in a long time. Assuming everyone is available on both weekends, who would you visit first? When a study posed this question, 90% said that they would get the unpleasant visit out of the way on the first weekend and see their friends on the second weekend, regardless of which option was presented to them first. Of course, if you visited your friends on the first weekend, you could revisit your fond memories of that weekend when the going gets tough on the second weekend; and yet, having something positive on the horizon was still overwhelmingly more appealing.

In another study, researchers purposefully raised participants' stress levels by asking them to prepare and give a five-minute speech about what makes them a good friend to others. One group of participants viewed a series of funny *New Yorker* cartoons before having to present the speech. Another group of participants were told that they would see the cartoons after the speech. The participants *anticipating* the positive experience felt better going into the speech than those who had just experienced it, and they felt less anxious when giving the speech. Having something enjoyable to look forward to was more helpful than having just had an enjoyable experience.

Having something to look forward to promises not only a positive future experience, but also an unknown that will become known. In the pilot episode of *Seinfeld,* Jerry is sitting on his couch, eating a bowl of cereal and watching TV, when the phone rings. He answers, "If-you-know-what-happened-in-the-Mets-game-don't-say-anything-I-taped-it. Hello." As a telephone greeting, this gets a laugh from the studio audience, and it turns out to be the wrong number. As Jerry is hanging up, Kramer makes his inaugural iconic entry, bursting through the door and yelling, "Boy, the Mets blew it tonight, huh?" Jerry falls to the floor in despair, not because the Mets lost, but because Kramer revealed the ending. Not knowing an outcome until it arrives is integral to the pleasure of many activities, and not just sports, books, and movies. One study found that people prefer to delay learning whether the lottery ticket they are holding is a winning ticket. While the future is uncertain, we get to experience the thrill of possibility. For as long as we can delay knowing that future, anything is possible.

In *Stumbling on Happiness*, Daniel Gilbert explains how uncertainty magnifies emotional experiences—both positive and negative. While the wait to find out the results of a medical test makes the experience even more unpleasant, the wait to find out which prize you won makes receiving that prize even more pleasant. We know this from a study conducted at the University of Virginia, where every participant was a lucky winner. Prizes included some chocolate, a bottle opener, a coffee mug, and other small items of similar value. Participants were asked to indicate their first-choice prize as well as their second choice. The first group then spun the wheel, learned whether they had won their first or second choice (depending on where the wheel landed), and then completed a series of other tasks. A second group first did the tasks, and were told that they would spin the wheel once those tasks were completed. A third group was told that since the study was almost over and because there were extra prizes, they wouldn't need to spin the wheel; instead, they would get *both* their first and second choice prizes. The researchers also asked the participants to fill out mood questionnaires at multiple points throughout the study. When they first learned about the wheel of prizes, everyone was generally feeling pretty good. But only the participants who didn't yet know which prize they would win still felt good at the end of the study—even better than the participants who were awarded both prizes.

Logically, two gifts should be better than one. But what the mystery gift holds that the two certain gifts do not is an *uncertainty*. There is joy in imagining all the different ways in which a future positive event will unfold.

The Good, the Bad, and the Savoring

As we look forward to future events, we spend more time thinking about them and what they might look like. And the more we think about and anticipate a positive event, the more we tend to appreciate it when it arrives.

Savoring is the psychological term for the actions you can take to increase your attention to and appreciation for positive experiences. You can savor events *before* they arrive by imagining them and looking forward to them—like looking forward to a weekend with friends and imagining the good times to come. You can savor events *afterward* through private reminiscing and through sharing with others—like my study participants who planned to share their good news and small wins. You also can savor experiences *as they occur*. If you search for stock photos of "savoring," you'll see several images of people so focused on some positive experience—the warm embrace of a sip of coffee, the delectable bite of food perfectly prepared, the smell of a flower, or the feeling of fresh air—that they close their eyes to sever visual distractors from the joy of the moment. Advice like "stop and smell the roses" encourages us to savor the good things in life, and the tendency to savor the good is associated with increased life satisfaction.

You might think that the reason we need advice like "stop and smell the roses" is that positive events are rare, and so it's important to appreciate them when they arrive. But we need this advice for the exact opposite reason. Positive events are actually much more common than negative ones. We often fail to realize this, however, because positive events often do not feel very unique.

A study asked participants to go about their days while carrying audio recorders, and found that the words people used were generally more positive than negative. And in a huge expansion on this idea, one research group amassed a corpus of more than 100 million words, captured from a wide range of sources including books, personal letters, and even diaries, and also audio recordings of conversations, speeches, and meetings. The data showed that people use positive words much more frequently than negative words. For example, "good" is used five times more frequently than "bad" (795 vs. 153 times per million words). Another study buzzed people every day for a week, ten times a day, to ask how they were feeling, and found that most of the time people feel good.

So, people speak more positively than negatively, and most of the time people feel good. So why does it often feel like there are more negatives than positives? To find out, one study asked participants, every day for a week, to write down one positive event and one negative event from their day. Then, at the end of the week, participants indicated how similar the week's events were to one another. The participants saw their past positive events as fairly similar to each other. And while the past negative events were seen as different from the positive events, the negative events were *not* seen as similar to one another. This is why negative events feel more newsworthy—they each feel unique, and therefore bad in their own individual way—and so negative events capture our attention and linger longer in our minds.

For something positive to stand out, it needs to be *really* positive. And so we will often go to great lengths to make ordinary positive events feel more special. This is where secrecy comes in.

GIFTS, PROPOSALS, AND OTHER SURPRISES

In 2013, two avid hockey fans, Joanna Chan and Julie Morris, embarked on a quest to see a game at each of the thirty-one NHL arenas. It took them five years to arrive at their final arena, the Saddledome in Calgary. To Morris's surprise, during the first period of the Calgary Flames game, with a microphone suddenly in hand, her partner got down on one knee and proposed. This was all displayed on a gigantic screen for the audience to see.

It turns out that there is an entire industry of proposals that take place on the jumbotron. The Saddledome charges on the higher side, at $5,000 (if you are looking for a bargain, you can propose on Fenway Park's screen for only $350), but Chan said that it was worth every penny.

"It's not about the public aspect of the proposal," but rather, "it was really just about making that last game special," Chan later said in an interview. And it was special. "I was completely overwhelmed. I was completely caught by surprise," Morris said. For a moment, Morris said that she forgot she was on the arena's big screen, "and then it quickly all came back when I heard the cheers, and everyone was all around us and excited, but there was a brief moment where it felt surprisingly intimate." She said yes, and they are now the Chan-Morrises.

The emotional experience of surprise is one of humans' most fleeting emotions. So why even bother at all? Surprises, by definition, contain some level of unexpectedness, and we tend to better remember unexpected happenings and treat them as more special. This helps explain extravagantly orchestrated marriage proposals like flash mobs, airplane banners, and, in one case, a TV commercial that the would-be groom created and had aired during his soon-to-be fiancée's favorite show. While the surprise itself may be fleeting, we can relive it vividly in our memories. This allows us to savor it long afterward.

Would you ever buy something just so someone else could tear it to shreds? Put that way, wrapping a gift sounds like a pointless and frivolous exercise, not to mention a colossal waste of often non-recyclable paper. But it does offer a psychological benefit. People have been wrapping and unwrapping gifts for more than a thousand years, according to anthropologist Chip Colwell. The paper itself elevates an object into a gift: even when you have a sense of what's inside, like a wrapped book, its exact nature begins as a mystery. Unwrapping a gift allows us to experience the thrill of anticipation and surprise; its contents remain a secret until the moment of revelation.

Whether it's a gift, a marriage proposal, a party, or anything else, to truly surprise someone, you need to do some planning. My wife, Rachel, loves surprises more than anyone I have ever met: both planning and receiving them. When we first met at a conference, Rachel lived in Australia and I lived in California for grad school. I visited Rachel in Australia a few months after the conference, and then later that year I planned to spend Christmas with her in Australia. Which is why I was dumbstruck when I got home one day, early in November, to find Rachel standing in the middle of my living room. Given that she *still had been in Australia* when we'd spoken less than 24 hours ago, the fact that she was in California, let alone in my apartment, seemed like such an impossibility that I refused to believe what my eyes were telling me. "Wow, that person looks *just* like Rachel," I thought as I stood there, staring at her, my face contorted—not because I was unhappy to see her, but because I was so very confused.

Rachel later told me that she had spent much of her 19-hour journey to Palo Alto imagining my joy and excitement upon seeing the huge surprise she had orchestrated with the help of my friends. But in the moment of revelation, she later told me, I looked kind of horrified, and she thought of the comment her SuperShuttle driver had made, as they chatted while awaiting another passenger: "You're completely surprising him? Wow. You know, not everyone likes surprises." The driver was right, in the sense that not everyone likes being *caught off guard*. Imagine you come home after a long day, thinking that you are about to have a nice quiet evening at home—and everyone you know is there and yelling "*Sur-*

prise!" That's more than surprising; that's putting you in the spotlight and thrusting you into a party you might not be in the mood for, with no notice. That might not be for everyone. Or take marriage proposals. While a surprise jumbotron proposal might be an unforgettable thrill for some, others would be unhappy, or even angry, to be put on the spot in that very public fashion. Catching someone off guard can backfire. Not discussing marriage in advance of a proposal, for example, is a major predictor of that proposal being rejected. Among rejected proposals, one study found, only 29% of the couples had discussed the idea of marriage in advance, whereas among accepted proposals, 100% of the couples had. So you might want to check in when it comes to certain surprises, and make sure you correctly understand the other person's attitudes toward marriage, surprise parties, extravagant gifts, and so forth—but otherwise, it's hard to go wrong.

Now imagine that instead of a surprise party, a life-changing proposal, or some other bombshell that catches you off guard, you received the perfect gift—something you absolutely love but had no inkling was coming. That sort of surprise is typically delightful, and delight is exactly what I felt once I understood that it really was Rachel standing in my living room that day.

Unlike the secrets discussed in the previous chapters, keeping something positive under wraps until just the right moment feels *good*—even if it requires careful monitoring, avoiding conversation topics, or hiding the evidence. The person planning the surprise is able to *savor* it from the moment the idea is conceived, and also gets to *look forward* to the revelation and the other person's (hopefully) delighted response.

For the last eight Christmas Eves, I've surveyed participants planning to give gifts the next morning. I've also conducted studies of secret good news like pregnancies, and other happy surprises. In total, more than 5,000 participants have told me about their positive secrets and what it's like to have them, and I find that the vast majority describe their experiences as vitalizing, exhilarating, exciting, and energizing. This is easy to see when we look at how people talk about their positive secrets.

> "I'm so filled with joy and energy right now. I feel like I can get so much done in the day and I think that's because of my excitement . . . In terms of keeping it a surprise to her, it's been a bit of a fun challenge."

> "Keeping it a surprise makes me feel like a secret agent on a top-secret mission. It's quite exhilarating and I cannot wait to see the look on her face! It makes me feel bursting with joy to wait until I unveil it for her."

While most people with positive secrets express pure joy and excitement, for some the anticipation is tinged with frustration, apprehension, and impatience.

> "I am very excited, but also very apprehensive at the same time . . . I am totally bursting to tell my colleagues, and it's

been a real struggle each day knowing what I know and not being able to celebrate my news."

"I will not tell anyone until [it's been] four months. It's like something is stored in my mind, and my head will explode."

Even in the face of challenge and apprehension, positive secrets can invigorate because we typically feel in control over the information and its delivery, including the setting, timing, and manner. In our studies of couples currently pregnant and keeping it secret (from at least some people), the more that pregnant women felt in control over who knew about the pregnancy, the more energized they felt by the secret, and the same was true for their partners. By the same token, it can feel quite frustrating or even burdensome to not have complete control over when we can share positive news—like in the case of winning an award or promotion and not being allowed to talk about it until it's official or publicly announced.

Feeling in control is one of the most important qualities of human life. When we feel in control of our lives, we feel more capable, and so we better manage stress and more effectively cope with life's challenges. Those who feel in control of their lives are happier and healthier than those who don't. They also live longer.

Some of life's most joyous and momentous occasions start off as secrets we keep in order to reveal: the gifts we give, pregnancies, marriage proposals, and more. Both the joy of anticipating the revelation of a positive secret and the feelings of control that come with the positive secret are experienced as satisfying and energizing.

SECRET JOYS

We've been discussing the ways in which secrecy can make a positive event all the more exciting when you reveal it. But there is another kind of positive secret, one that you may have no intention to ever reveal. I call these secret joys.

Let's hear from two of my research participants describing the joyful experiences they keep to themselves.

"I enjoy meditating early in the morning, but I don't tell other people about it. Their reactions might spoil my meditations. I would not want to remember their comments while spending my quiet time meditating and reflecting."

"I love going for walks at nighttime ... I find the experience peaceful because my neighborhood is whisper silent, and the glow of the sodium-light street lamps just puts me at ease. I do not tell other people I like going out at night because I know they'll question how safe it is and try to prevent me from doing it. I feel safe and never in danger. I need that time to myself [which helps me] unwind, let go of my stress, and find solace in myself."

Compared to gifts, proposals, and other surprises, there is a different tone to these positive secrets. Rather than exhila-

POSITIVE SECRETS | 171

ration and bursts of energy, these are stories of contentment and independence.

When it comes to joyful experiences we don't tell others about, it becomes easy to see how there are other flavors of solitude than isolation. Rather than feeling alone, you may feel independent and free from others' expectations and opinions.

Many of the hobbies that people keep secret fit this mold. These include artistic pursuits, collecting (cards, comics, stamps, coins), gambling, needlework (crochet, knitting, etc.), meditating, games (card, board, video), reading (e.g., sci-fi, romance novels), recreational drugs, working out, writing (e.g., fiction, poetry), watching TV (e.g., cartoons, children's shows, soap operas), and yoga.

All activities, of course, carry some risks—recreational drugs, walking home alone late at night, watching a TV show into its seventh season—but when my participants tell me about their secret joys, they tell me that they know the risks and have considered them carefully. Part of their enjoyment of these activities is experiencing them without having to deal with people who would be judgmental about the activity.

What distinguishes secret joys from surprises like gifts, proposals, and good news is that *positive surprises* amplify a positive experience, whereas *secret joys* are kept to protect the positive experience.

We tend to think about coping in the context of negative life events. And yet, positive life events sometimes call for coping too. Imagine, say, you just won a video game; what would you do? You might tell someone else, hoping the other person will respond positively, expressing appreciation for your accomplishment or giving congratulations. But what if the person responds poorly instead? Have you ever been excited to share something with someone, only to receive a "so what?" response? I certainly have, and it can be crushing. When someone responds negatively to something you enjoy, it can leave you feeling worse, wishing you never opened up. And so when people enjoy something and know that other folks would disapprove or not understand, they might keep it to themselves.

Rather than isolation, these secrets can provide healthy forms of solitude: independence and autonomy. Sometimes you don't need others' thoughts on a matter, and it helps to recognize when that's part of your decision. With positive secrets, it is clear that we are in the driver's seat, either cruising along toward a destination we've been looking forward to, or taking a joyride to a place that is special and just for us.

We have a degree of control over all our secrets. We choose which ones we share, and how we share them. But for many of the secrets we hold back, it may not seem clear to what end we exercise this control that we have—especially as we try to reconcile our secrets with our human needs to relate, connect, and be known. Understanding these universal and human desires provides us with one final window into our secrets.

CHAPTER 8

Culture and Coping

E ach time Hong Lu collected the test result, the conclusion was the same. When the third prognosis—stage four
lung cancer—confirmed the prior two, it was undeniable.
She decided, then and there, that her sister could never know,
and so she intercepted the medical report and went to a print
shop. She asked if they would help her create a medical report, but they refused. Creating false documents clashed with
their code of ethics. But if she were to provide them with a
document, they said, they would be willing to photocopy it.
And so, with a few strokes of a tiny brush, she hid the truth
underneath a layer of opaque white. In the new blank space,
she wrote a new diagnosis, which sounded medical but be-
nign. She made a photocopy of her doctored document. Her
sister—Nai Nai, as we'll call her—had been given three
months to live, and her family was determined to keep this
secret from her.

Lulu Wang was born in Beijing and moved to the United

States when she was six. A love for storytelling led her to study film production at Boston College, and her second feature-length film shared the story of her grandmother, Nai Nai, and the secret her family had gone to great lengths to conceal, as Wang first recounted on NPR's *This American Life*. "We realized we had a problem. How are we all going to manage to see Nai Nai before she died? Family from three countries needed to say goodbye without letting Nai Nai know we were actually saying goodbye. This would require more than Wite-Out." Luckily, Wang's cousin had a wedding celebration planned, so the family decided to put it on one year early. The entire family convened, giving everyone a chance to see the family matriarch before her passing. And all the while, she remained completely unaware that her family members were saying their goodbyes, unaware of her terminal cancer diagnosis, and unaware that her doctors estimated she only had three months left to live.

"When I heard the news, I wanted to talk with my grandmother, comfort her. I wanted to grieve with her in the way that seems natural when someone you love is dying." But Wang was forbidden from informing her grandmother about the diagnosis. If she wanted to see her grandmother one last time, she was told, she would have to participate in the elaborate deception.

Wang explains that it is customary for medical doctors in China to give bad news to family members, especially when it comes to older patients. This is what allowed Wang's family to intercept the news of the diagnosis and keep it secret. Wang's film adaption of her story, *The Farewell*, stays close to the real story, filmed on location and even with Nai Nai's real sister, the one who falsified the medical report, playing herself in the movie. The story fascinates the audience unaccustomed to this kind of secrecy. This simply could not happen in many places, at least not nearly as easily.

While she joined her family in keeping the secret from her grandmother, this was not Wang's first choice. If she had it her way, she would have told Nai Nai. But Wang prioritized the wishes of others over her own; she kept a collectively held secret that she did not want to keep. While the secret did not concern her personal affairs, she still felt shackled by it. She could not freely engage with her grandmother the way she so desired. She had to lie to Nai Nai about something huge, and it felt wrong. Her experience of participating in this secret highlights how our experiences of secrecy are shaped by our culture.

Like the air we breathe, culture is all around us, touching and affecting every fiber of our social universe. And so, of course, culture—inherited from our families, neighbors, friends, coworkers—will play a role in how we relate to our secrets. Did Nai Nai have a right to know the truth? Straddled across two cultures offering different answers to this moral dilemma, Wang felt conflicted. She wanted to respect the collective decision made by her family, and understood their desire to protect her grandmother, but she also felt that her grandmother was owed the truth.

"I owe someone the truth about this." In one study, I asked more than 7,000 participants, from twenty-six different countries around the globe, to answer a simple *yes* or *no* to this very question, about each secret they had from our list of 38. I found that, of the nearly 80,000 secrets they kept in total, for 20% of them, they felt they owed someone the truth. Do we owe others the truth when it comes to our secrets? The answer to this question differs from culture to culture.

The air metaphor for culture is useful, because it serves as a reminder that while culture is a powerful force in aggregate, in any particular situation or moment in time, it may only have a light touch. My participants' home countries contributed to how they experienced those 80,000 secrets, but not to the extent you might think. I find that the *category* of secret you are keeping (of the 38 from our list) has ten times more influence on how you experience your secret than which country you are from; and your *personal disposition* has more than twice that (twenty-eight times more influence than which country you are from).

And so, our experiences of secrecy are much more similar to those of people who have the same types of secrets or similar dispositions, rather than random people from the same country. But my global study of secrecy didn't just look at the country my participants were from. It also looked at specific features of the environment they inhabited; for example: Is it easy or difficult to form new relationships? Is your social network clumped into a tight cluster like a constellation packed full of stars, or does it look more like a galaxy of solar systems with their own internal orbits, to which you belong and travel through? And then, for the social group in which you are most embedded, would you sacrifice your own interest for its benefit? Would you put the group before yourself? Your answers to these questions have direct implications for whether you tend to keep secrets or reveal them, and the emotional experiences that come with each.

MANAGING RELATIONSHIPS

In the summer of 2012, backpack in hand, I arrived alone at a hostel in the Sultanahmet neighborhood of Istanbul. After depositing my belongings in my bunk, I decided to hit the rooftop bar. Scanning the area for friendly faces, I approached a group of similarly aged travelers to ask if a seat was taken. They invited me to join them. As someone who is not especially outgoing, I was surprised by how easy it was to strike up a conversation and make quick friends. This turned out to be the case in every city I visited, and after just two weeks, I was socially exhausted from chatting with all the new people I had met.

The environment that a hostel offers, with its shared rooms for dining and hanging out, is particularly conducive to these kinds of friendly conversations and connections. And the people who opt to stay in hostels are more likely to be traveling alone, and to not mind spending time with strangers; for both of these reasons, they are typically inclined toward connecting with new people.

Some environments allow more opportunity for forming relationships than others. This is what psychologists call relational mobility. Is it easy to meet people in your environment? Is it normal for people around you to have an extended conversation with someone they've just met? Can you easily discard friends, dissolve relationships, and replace them with new ones? Relational mobility varies in degrees, and is a feature of any environment that people inhabit, whether it's a youth hostel in Istanbul, your hometown, or where you currently reside.

In low relational mobility environments, bonds are not easily broken. In some regions, the idea that you could lose touch with your parents or break off from family is unheard of; such bonds are considered permanent. And divorce might be difficult to obtain in such environments. At the extreme end of low relational mobility, you may have less choice over the person you marry in the first place, not to mention the inability to split up if things don't work out.

Spouses, friends, and acquaintances in low relational mobility environments are concentrated within smaller social networks, such as your hometown, and it is difficult to form relationships outside of these tight-knit networks. In contrast, in high relational mobility environments, relationships are formed through personal choices and traveling through multiple social spheres. People search for each other in the marketplace of potential relationships. In these environments, people tend to be more trusting of strangers and therefore more inclined to approach and connect with new people.

In general, East and Southeast Asia, Northern Africa, and Arabic-speaking countries tend to be relatively lower in relational mobility, whereas North and South America, Australia, and Europe tend to be relatively higher in relational mobility. Of course, these are just trends; each country contains a diversity of individuals and cultures mixing and mashing. And so, relational mobility varies within countries too. Cities, for instance, offer far more relationship opportunities than do small towns.

Certain periods of life also offer more opportunity for relationship partners. Moving into a college dormitory, for example, offers ample opportunity for developing relationships outside your existing social circle—exactly what the hostel bar offers for solo travelers.

Importantly, opportunity for social connection does not automatically translate to obtaining it. I had to put myself out there on that rooftop to make my Istanbul friends. Going up to strangers brings some degree of risk; they could have easily rejected me. But had I not connected with them, I might have felt lonely, looking at all the other people immersed in their conversations. And so, with higher relational mobility, people tend to take more social risks: not just going up to strangers, but also expressing dislike for something people tend to like, and even making life-changing decisions like moving to a new location where you don't know anyone.

Relational mobility, as a feature of your environment, is neither good nor bad. A significant downside of high relational mobility is that it makes it easier for people to reject you. Your friends and romantic partners will have an easier time ditching you for someone else in high relational mobility contexts, and so if you want to maintain the relationship, you'll have to put in more work. In high relational mobility environments, romantic relationships are more passionate, friendships are more intimate, and here's the big one: people self-disclose more. It takes trust and courage to open up, even with people you are close to. Making yourself vulnerable in this way is an act of intimacy and a strong signal of commitment in high relational mobility contexts.

So what does this mean for our secrets? In my study of those 80,000 secrets, I also asked my participants how much relational mobility their environments offered. Across the twenty-six countries I've surveyed, I find that people in lower relational mobility environments tend to keep more secrets, and also tend to feel more isolated and alone with those secrets.

When we cannot easily move from relationship to relationship, we tend to take fewer risks in what we disclose and share. And if the people around you do not share weighty secrets with one another, you may feel that there is no world in which you can reveal your secret. Understandably, this would feel isolating. This makes low relational mobility sound like an added burden when it comes to secrecy, but that's not necessarily true.

If your environment is *not* one in which people often connect with one another by revealing their innermost thoughts, then when you keep a secret, you are less likely to feel that you are violating some norm or rule of the relationship; and so you are less likely to feel guilty. In higher relational mobility contexts, while people feel less isolated with their secrets, they feel more guilty for having them.

MANAGING VALUES

"For me, it was a really big secret. I wasn't comfortable keeping it, but I knew I had no choice. Grandma played a huge role in this. She wasn't ever going to be comfortable with [revealing] it, never." This is my mother describing her inner conflict about keeping the secret that my brother and I were donor-conceived. Had it been up to her, she might have revealed the secret sooner, but our grandmother was worried that my brother and I might not feel part of the family if we were ever to find out. "I was caught in a very awkward position because I was forced to keep a secret that I really didn't want to keep forever." Her experience of prioritizing the wishes of others over her own echoes Lulu Wang's experience of keeping her grandmother's diagnosis a secret. Wang was conflicted between the value system of her family, and what she felt was very wrong—lying to her grandmother. This left her with only questions, and no clear answers. "If I talked to my American friends, [they] would have the American answer, and then my family would have the Chinese answer," she recalls. "And so you just felt very torn." Though my mother wasn't torn between two cultural norms, she was torn between two desires: to tell us the truth, and for us to feel part of the family as much as we always had before.

In Wang's film, her character's uncle says, "You want to tell Nai Nai the truth because you're afraid to take responsibility for her. Because it's too big of a burden. If you tell her, then you don't need to feel guilty. We're not telling Nai Nai because it's our duty to carry this emotional burden for her."

Behaviors connected to duty, obligation, and putting others before yourself are hallmarks of collectivism, a broad cultural construct characterized by interdependence, a focus on others as opposed to the self, collective goals, and values for group harmony and cohesion. Collectivism occurs within many different types of groups—a small residential or religious community, a workplace, or another institution—but where collectivism really shows its stripes is within family dynamics. We all belong to several groups, and sometimes a group's interest conflicts with our own.

Prioritizing the group over individual interests is certainly a universal, cross-cultural phenomenon; however, collectivism is stronger in some cultures than it is in others. And as before, even in the same country there will be a mix of cultures and individuals. For example, one study asked participants—from the Philippines, the United States, and Turkey—how religious they were, and to what extent they endorsed collectivistic values. The more the participants were religious, regardless of their nationality, the more they were oriented toward collectivism.

"I am prepared to sacrifice my own interest for the benefit of my group." "I think it is more important to give priority to group interests rather than to personal ones." Folks who agree with statements like these tend to give priority to collective goals over personal goals when they conflict. I asked my participants to indicate their agreement with statements like these as a measure of the extent to which they were oriented toward collectivism.

While *low* relational mobility is associated with less self-disclosure and more secret-keeping, collectivism does not uniquely relate to the amount of self-disclosure or secret-keeping that people engage in. Rather, collectivism shapes how people experience the secrets they do have. Even if keeping the secret protects group harmony, secret-keepers concerned about the group act as if others deserve the truth about their secret, and feel more inauthentic and ashamed for having the secret.

So while collectivism is not associated with having secrets that are more immoral, it is associated with *the experience* most closely linked with having immoral secrets: shame. Holding back from others itself feels more wrong. Similarly, collectivism is not associated with having secrets that are more relational, yet it is associated with the experience most closely linked with having relational secrets: feeling inauthentic.

Recall our map of secrets, and how the relational dimension ranges from individually oriented secrets (not involving others) to highly relational secrets (very much involving others). A detail I've saved for now is that the more a secret is relational, the more we tend to feel inauthentic for having it, as if we are holding back some core part of ourselves from others.

Who you are is not fully separable from your relationships. And when our relationships sit within more collectivism-oriented communities, to have a secret seems more in conflict with other values. But that doesn't mean keeping the secret is wrong. In many cases, we keep secrets with the intention of protecting others or our relationships with them. In a study where we gave participants a coping compass that pointed them to this very recognition—of relational benefits and prosocial aspects of their secrecy—they felt more authentic as a result.

Did Nai Nai deserve to know the truth about her cancer diagnosis? On the one hand, you can imagine wanting to know about this so you could spend your final months thoughtfully, say your goodbyes, and perhaps have the opportunity to die without regrets; but on the other hand, it's hard to imagine a truth that would be more difficult to confront, so perhaps ignorance is bliss. But what if I told you that, in the end, Nai Nai lived beyond those three months? That, more than seven years after her diagnosis, she is still alive, and the family attributes this happy outcome to keeping the secret? "The Chinese believe that mental and emotional health are completely linked to physical health," Wang told *This American Life*. And according to Hong Lu, the reason her big sister survived against all odds was "because we gave Nai Nai joy instead of worry." There is no obvious right answer here, but perhaps in this case keeping the monumental secret was, on balance, a positive.

When it's unclear whether a secret is better off revealed or concealed, don't focus only on what you are holding back. Reflect also on what you are putting forward, whether it is the emotional health of someone you care about, group harmony, or something else. If you are keeping the secret with someone else's best interests at heart, even when you feel conflicted about it, recognize you are putting the needs of others before your own. There is nothing inauthentic about that.

MANAGING EMOTIONS IN TIMES OF NEED

When I first started graduate school, I studied how we perceive emotions on others' faces. Even when expressions are subtle, we often can tell when someone is happy, angry, surprised, or sad. And it's not just these basic emotions that we can see on others' faces. We can also tell when someone looks skeptical, for instance, or uncertain or determined. Our inner feelings are on the inside, and yet they are often *also* visible on the outside. Why?

Unless you are looking at your own reflection, your facial expressions are not visible to you at all, but they are to others. The fact that facial expressions of emotion serve to communicate our feelings to others has long been established (Charles Darwin wrote a whole book on the subject), but only recently have we begun to understand the role of culture in emotion expression.

Much of the time we show our emotions on our faces, but we have some control over what we show, and sometimes it's more polite to mask a feeling than to express it. An eye roll, for example, is rarely well received (at least in my experience). And so, whether you express an emotion may depend on how well you think it will land. Here also, culture seeps in. The more you are concerned with group harmony, the more you hold back expressions of emotion that would disrupt the harmony. This doesn't mean dialing down the feeling, but rather, keeping it to yourself.

Those oriented toward personal independence place more importance on expressing feelings to others and asserting them as valid. This influence extends beyond emotion expression to self-expression more broadly. At one extreme, some cultures value speaking your mind as an important form of self-expression; and at the other extreme, other cultures may view silence as a sign of respect and attentiveness, and speaking with others is more about maintaining relationships than expressing oneself.

We've discussed how culture has sway over how we experience our secrets, but what about the *content* of our secrets? The only cultural difference we see across the three dimensions of secrets is that collectivism is associated with keeping more emotionally based (and thus less goal-oriented) secrets. This lines up with cultural differences seen in emotion expression, where those in collectivistic cultures more frequently hold back emotion expression that could disrupt group harmony.

If there is any silver lining when it comes to secrets about your past, it is that nobody can see those secrets on your face. But this does come with a cost of its own. If there is no easy way for people to tell when you have a hidden inner struggle, then it is on *you* to find or ask for the help you need.

I've been extolling the benefits of confiding secrets and receiving social support throughout this book, but the idea of explicitly asking for help with personal problems is an added stressor for those who are already concerned about the negative impacts a disclosure could have on the relationship. These relational concerns explain a cultural difference seen in the tendency to actively seek social support.

Heejung Kim, a professor of psychology at the University of California, Santa Barbara, has extensively studied cultural differences in social support-seeking. She finds that in Asian cultures, there is a reduced tendency to explicitly ask close others for support. This doesn't mean that individuals concerned with group harmony do not benefit from social support. Rather, support—both how it's sought and how it's provided—follows cultural norms and values. In cultures that prioritize the well-being of the group over the individual, people are more concerned about placing burdens on others, and so are more reluctant to ask for help with their own personal problems.

Kim and her colleagues asked Korean and American participants to report their most stressful event of the day, every day for a week. Each day, the participants reported how many people they interacted with after the stressor, whether they spoke about the stressor, the emotions they experienced that day, and daily life satisfaction. The researchers examined two kinds of social support: talking about the stressor with others, and spending time with others but without talking about the stressor itself. For example, if you are having a problem at work, you might benefit from talking about it with others, whereas if you're dealing with a breakup, you may not want to talk about it, but you may not want to be alone with it either. So, each kind of support has its value. But does this differ from culture to culture?

Kim's study found that American participants spoke with more people about their stressor, whereas the Korean participants spent more time with others, but without talking about the stressor. While Korean participants benefited from both kinds of social support, asking for support was associated with feelings of shame for having burdened another person with one's own problems. Americans, in contrast, benefited mostly from talking about their stressor, and they did not feel bad for having brought their problems to another person.

Remember that culture shapes our experiences of situations like these, but it doesn't define them. For example, in open-ended responses from one study, Kim found that 39% of Korean participants reported seeking social support to relieve stress, relative to 57% of her American participants. Another study asked the extent to which participants sought various kinds of social support, and found—on a scale that ranged from 1 (no support seeking) to 5 (extensive support seeking)—Asian immigrants to the United States were just above the midpoint of the scale at 3.3, U.S.-born Asian Americans were nearby at 3.5, and European Americans were at 3.9, just a tad higher. And so, the Americans sought social support a bit more, but the differences here are small. While the nature of support-seeking and support-providing may look different from culture to culture, the tendency to value social support and to benefit from it is universal, when it comes in the right form.

If you feel isolated with a secret, know that this might signal a need to talk with someone about it. If you feel inauthentic for having a secret, know that this might signal some inner conflict that is worth identifying. Whatever your struggle, know that everyone benefits from social support. When it comes to coping with personal struggles, we have more in common than we do differences.

CHAPTER 9

Secrets Shared

We've heard from lawyers, a whistleblower, a mob boss, teenagers dealing with worries and insecurities, a person with a decades-old secret that would never have been discovered, and thousands of research participants. They all suffered with their secrets. Yet their suffering rarely came from the work it took to keep their secrets hidden, but rather, from having to carry the secrets alone without the support of others. Holding back some part of your inner world from others can be associated with feelings of shame, isolation, inauthenticity, and uncertainty. But it doesn't have to be that way. Not all secrets are negative, and even the ones that are don't have to have the power over you that they often do.

We can share our secrets with others. We've heard from people willing to reveal their secrets on the Secret Telephone, on mail-in postcards, and to strangers they just met: revelations that came with no costs. But you need not search so far and wide to confide; family members, close friends, a romantic partner, a colleague—they also make for great confidants, people who can see the whole picture and who won't forget everything else they know about you when learning something new. People who will empathize, and forgive.

At the end of every study I conduct, I include a blank space for participants to leave any comments they wish to share about the study and their experience. People often express surprise about the value they found in taking a hard look at their secrets. Whereas some secrets can take hold of your thoughts and demand your attention, others will recede into the background, nearly forgotten. Secrets can get easier with time, but you can also take a more proactive role in the process. You don't have to wait for years to pass by.

If you take away only one lesson from this book, I hope it is this: If you have a secret that is bothering you, consider sharing it with someone you trust. If there are legal or professional reasons for the secrecy, then your situation is more complicated, but there still may be *someone* you can confide in. Even if you don't reveal the specifics, you can still benefit from talking to others. Rather than turn inward, look outward, and seek others' help.

A colleague once told me a story about a woman who was seeking help. She was trying to locate her biological relatives, and she found a match through a genetic testing service. It turned out that she had found her half-sister, who could confirm that her father had been a sperm donor in the past. Over the internet, she found her biological father.

If you are donor-conceived, you can't hear a story like that without wondering: Could I too find a match out there? I decided to order a genetic testing kit. I wasn't sure what I was looking to learn when it arrived, but I sent in my sample. And I did this with some hesitation and some secrecy. Some of my friends knew, but not all of them, and I didn't tell *anyone* in my family. I just wasn't ready. I wanted to know the full story before sharing it with others. I wanted some time to have it to myself, to process it, and to find my footing.

If you are not ready to reveal your secret to someone, ask yourself why that is the case, and also remind yourself why you are keeping the secret in the first place. Importantly, could someone learn it without your telling? And are there people in your life who would expect you to share this with them? Your answers to these questions will point you to your best path forward.

If the secret drags on your well-being, even just occasionally, take a close look at that hurt, and try your best to understand it. Now consider that perhaps keeping the secret does not cause harm to anyone else. Perhaps your secret protects someone you care about. Or maybe you have your reasons for keeping the secret. If one of these seems closer to your situation, then that is your coping resource, your lifeline.

We share our inner worlds with others to form and maintain relationships, but sometimes we hold back in order to protect those relationships as well. In other words, we *don't* share for the same reasons we *do:* we seek to establish and maintain close bonds.

Besides a shared experience or physical touch, sharing our inner world with others is the central way to connect with others and to be known. And sharing a secret with someone— something that you wouldn't tell just anyone—is a profound act of intimacy. Opening up, whether in a newly formed relationship or in a long-established one, brings you closer to your confidants and deepens your relationships with them.

I asked my mother when she first started to consider revealing the family secret, and I had to laugh at her response. "When did your first paper on secrecy come out?" It was my own research that had changed her thinking. "I felt like it became even more important than ever to not have this secret because I was reading your research and what you wrote, and I said, wow, this is really not a good thing . . . I started to think how important it would be to finally communicate." She then told me a story about how she has since freed herself of more secrets. She talked about how finally sharing things that she had been keeping from others improved her relationships instead of hurting them. Now she feels more supported by and connected with others. During the course of our conversation about secrets, my mother shared an additional secret with me, and I was touched that she felt comfortable doing so. The conversation made us closer.

In a way, we've known this all along. Children are well aware of the positive social power of secrets; even young children associate secrets with a sense of intimacy. Children will tell you that a secret is something you would only tell your best friend. Once they learn that they have an inner world, known only to them unless shared, children selectively share that inner world with others. Kids get it, but as adults we too often forget it. *Secrets are meant to be shared.*

A rich, private inner world is a wonderful thing. It is where we fondly recall past events and fantasize about potential futures. And you have control over who you let into your inner world. When we choose to share our inner worlds with others, doing so brings us all a little closer together.

———

A month after I sent in my genetic testing kit, the results came back. I found a match, but the information was sparse. Where a first name could have been listed was only an initial and a last name, followed by some numbers. So, no first name. Where a photo could have been uploaded was a blank square filled with the default headshot icon. So, no photo. Whoever this person was, all I knew was that we were closely related. If I wanted to learn more, I would have to send a message. It took a few days to work up the courage, and it was exceptionally difficult to find the right words. With the help of a few others, I wrote my delicately worded note, asked my delicately worded questions, and sent it in.

Two years passed, and I still hadn't heard back, and so I figured that this journey had hit a dead end.

And then, out of the blue, as I was working on my final draft of this book, I received a message from Alexis; and, not long after, a message from Ross (who, it turned out, was the person I had first tried to contact). Both of them—through the same genetic testing service—had just learned something startling, something that my younger brother and I had known about ourselves for years: they were donor-conceived.

A decade ago, I learned my family secret. And its implications continue to be revealed: two new half-siblings (we were all born within one year of each other) living on opposite sides of the country. We have since connected and traded stories. We have bonded over what it's like to learn such a major secret.

The stories that Alexis and Ross shared with me, of how they took in this information and how it has affected them, sounded surprisingly familiar, similar to how I first reacted to learning my family's secret. You can't help but wonder if the similarities in our stories have anything to do with us being half-siblings. It's been a lovely surprise to get a chance to know these two people with whom I share one of life's most fundamental bonds. Rather than creating wedges, some secrets—when shared—can bring people together.

ACKNOWLEDGMENTS

The people who are not you have this amazing quality of thinking differently than you do. Even if your best friend, your closest confidant, or your partner thinks in very similar ways to you, still they have fresh perspectives that they can offer you that are incredibly difficult to find on your own. A simple conversation with a trusted other so often makes a world of difference.

This book has benefited tremendously from conversations with others. I'm grateful to so many who've given me their time, their ears, and their feedback. Often, I used the word "we" when describing research findings, and that's because several of the studies I describe in this book benefited from conversations and collaborations with people that I'm lucky to call both my colleagues and friends: Katie Greenaway, E. J. Masicampo, Nick Camp, Adam Galinsky, Maila Mason, Sheena Iyengar, Nir Halevy, Alex Koch, Elise Kalokerinos, James Kirby, Brock Bastian, and Jessica Salerno.

Nalini Ambady was the earliest supporter of this research when I first started it at Tufts, and other folks who were instrumental in early stages of this research were Negin Toosi, J. S. Chun, Kevin Tayebi, and Adrien Aufort, who lugged water bottles and surveys all over Central Park in the sweltering heat, and fearlessly asked strangers to reveal their secrets. This research has also benefited greatly from conversations over the years with my colleagues: Modupe Akinola, Daniel Ames, Joel Brockner, Shai Davidai, Adam Galinsky, Tory Higgins, Sheena Iyengar, Malia Mason, Sandra Matz, Michael Morris, and Kathy Phillips.

I'm beyond grateful for my friends and colleagues who read this book in all its stages, and I'm lucky to have their support, and to draw from their perspectives, observations, and insights: Nick Camp, Shai Davidai, Alan Gordon, Katie Greenaway, Sarah Gripshover, Eric Hehman, Lauren Jackman, James Kirby, Ashley Martin, E. J. Masicampo, Sandra Matz, Rachel McDonald, Anna Merritt, Becca Neel, Dave Paunesku, Xenia Shih Bion, Dean Weesner, and Ashley Wright.

A special thank-you also goes out to friends and colleagues who stepped in to carefully consider and evaluate translations of the 38 categories of secrets (among other questions about secrecy) across more than a dozen languages, and with urgency, when I decided midway through writing this book to conduct a cross-cultural study on secrecy. It all came together just in time, thanks to Aleksandra Cichocka, Shai Davidai, Fabio Fasoli, Frederico Guilherme, Seval Gündemir, Alex Koch, Alice Lee, Ana Leite, Ioana Medrea, Benoît Monin, Julien Monteil, Élika Ortega, Petros Perselis, Takuya Sawaoka, and Ricardo Shih Bion.

This book would not be possible without the amazing help and support from the folks at Crown, including my editors, Gillian Blake, Emma Berry, Caroline Wray, and Talia Krohn. A special thank you goes to Talia, who provided feedback on the whole book through multiple iterations (including when deadlines were tight), and my agent, Margo Beth Fleming, who was there from the book's Day One. I was advised to meet with multiple agents before picking one, but I knew immediately after I met Margo that I would not be following that advice. Margo understood my vision early in the process, and helped me articulate it and refine it.

Thanks also to Sheena Iyengar for generously sharing her story, and my parents, Don Slepian and Judy Finazzo, for sharing their stories too, and for being willing participants, when it came to the new questions I only thought to ask when writing this book (which, after learning so much, made me wonder why I didn't ask these questions sooner; but this is what happens when we don't talk about secrets, even the ones that are out).

In January 2012, midway through grad school, while at a conference in San Diego, I sat down on a couch at a party, and struck up a conversation with the person sitting next to me. This will forever be the best decision I ever made because that person, Rachel McDonald, would be the person I married three and a half years later. Rachel is also a social psychologist, but we don't talk psychology at the dinner table. In fact, it was *not* wanting to engage in academic talk at a party that got our first conversation started. But whenever I'm stuck on something, I bring it to her, and she instantly has all the answers and the best ideas for how to move forward. No single chapter of this book saw the light of day or another set of eyes until Rachel saw it and gave me feedback first. I aspire to plan surprises that are in the same league as the ones she so wonderfully creates for others and myself, but that hill is a steep one. Rachel has been, and always will be, the best surprise of my life. I simply can't find the words to convey how lucky I feel to have her in my life, to have someone so thoughtful, kind, generous, funny, smart, supportive, and adventurous by my side. No words could do it justice.

NOTES

CHAPTER 1: WHAT IS A SECRET?

4   *"Nothing is harder"* Snowden, E. (2019). *Permanent Record* (p. 241). Pan Macmillan.

4   *"They could just spy"* Snowden (2019), p. 5.

5   *"The way to reveal"* Snowden (2019), p. 242.

5   *"This required"* Snowden (2019), p. 242.

5   *"I knew that disclosing"* Snowden (2019), p. 242.

5   *"I'd be sweating"* Snowden (2019), p. 259.

5   *"In other attempts"* Snowden (2019), p. 258.

6   *"I contacted the journalists"* Snowden (2019), p. 250.

6   *"You can't really appreciate how hard"* Snowden (2019), p. 251.

6   *"Not wishing to cause"* Snowden (2019), p. 241.

6   *"Hadn't I gotten used"* Snowden (2019), p. 241.

6   *"At least you're part"* Snowden (2019), p. 241.

8   *Neuroimaging studies find* McNorgan, C. (2012). A meta-analytic review of multisensory imagery identifies the neural correlates of modality-specific and modality-general imagery. *Frontiers in Human Neuroscience, 6,* 285.

9   *Interested in this physical* Slepian, M. L., Masicampo, E. J., Toosi, N. R., & Ambady, N. (2012). The physical burdens of secrecy. *Journal of Experimental Psychology: General, 141,* 619–624.

9   *When people feel fatigued* Proffitt, D. R. (2006). Embodied perception and the economy of action. *Perspectives on Psychological Science, 1,* 110–122; Witt, J. K., Proffitt, D. R., & Epstein, W. (2004). Perceiving distance: A role of effort and intent. *Perception, 33,* 577–590.

10 *If they said yes* Slepian et al. (2012).

11 *"big" versus "small" secrets* LeBel, E. P., & Wilbur, C. J. (2014). Big secrets do not necessarily cause hills to appear steeper. *Psychonomic Bulletin & Review, 21,* 696–700.

11 *When we ran our original* Slepian, M. L., Camp, N. P., & Masicampo, E. J. (2015). Exploring the secrecy burden: Secrets, preoccupation, and perceptual judgments. *Journal of Experimental Psychology: General, 144,* e31–e42.

12 *We ran the hill slant study again* Slepian et al. (2015).

14 *a survey I conducted* Slepian, M. L., Chun, J. S., & Mason, M. F. (2017). The experience of secrecy. *Journal of Personality and Social Psychology, 113,* 1–33.

16 *On average, they indicate* Slepian et al. (2017).

18 *"I've looked on a lot of women"* Scheer, R. (November 1976). "The Playboy Interview: Jimmy Carter." *Playboy* 23(11), 63–86.

23 *but we tend not to see* Baxter, L. A., & Wilmot, W. W. (1985). Taboo topics in close relationships. *Journal of Social and Personal Relationships, 2,* 253–269.

23 *you are trying to avoid* Sun, K. Q., & Slepian, M. L. (2020). The conversations we seek to avoid. *Organizational Behavior and Human Decision Processes, 160,* 87–105.

23 *not wanting other people* McDonald, R. I., Salerno, J. M., Greenaway, K. H., & Slepian, M. L. (2020). Motivated secrecy: Politics, relationships, and regrets. *Motivation Science, 6,* 61–78.

CHAPTER 2: THE BIRTH OF SECRETS

26 *we are able to understand* Epley, N. (2014). *Mindwise: How We Understand What Others Think, Believe, Feel, and Want.* Vintage.

28 *there are politics within* Goodall, J. (2010). *Through a Window: My Years with the Chimpanzees of Gombe.* Mariner Books.

28 *For example, if a chimp* De Waal, F.B.M. (1986), Deception in the natural communication of chimpanzees. In R. W. Mitchell & N. S.

Thompson (eds.), *Deception: Perspectives on Human and Nonhuman Deceit* (pp. 221–244; see p. 228). State University of New York Press.

29 *To find out, a research team* Hare, B., Call, J., Agnetta, B., & Tomasello, M. (2000). Chimpanzees know what conspecifics do and do not see. *Animal Behaviour, 59,* 771–785.

29 *In his book* De Waal, F. (2000). *Chimpanzee Politics: Power and Sex Among Apes,* Revised Edition (p. 37). Johns Hopkins University Press.

29 *Mating chimpanzees* De Waal (2000).

30 *When human researchers act* Melis, A. P., Call, J., & Tomasello, M. (2006). Chimpanzees (Pan troglodytes) conceal visual and auditory information from others. *Journal of Comparative Psychology, 120,* 154–162.

30 *The test goes like this* Hare, B., Call, J., & Tomasello, M. (2001). Do chimpanzees know what conspecifics know? *Animal Behaviour, 61,* 139–151.

33 *The infants look longer* Onishi, K. H., & Baillargeon, R. (2005). Do 15-month-old infants understand false beliefs? *Science, 308,* 255–258.

33 *A study with babies* Buttelmann, D., Carpenter, M., & Tomasello, M. (2009). Eighteen-month-old infants show false belief understanding in an active helping paradigm. *Cognition, 112,* 337–342.

34 *Up to a certain age* For a summary of the similarities, and where children surpass chimps, see Call & Tomasello (2008). Does the chimpanzee have a theory of mind? 30 years later. *Trends in Cognitive Sciences, 12,* 187–192.

35 *For instance, a parent* The parents' stories that I share here come from a study where I asked parents of children of all ages to tell me about times their children shared secrets with them and attempted to keep secrets from them.

36 *If an alpha male catches* De Waal, F. (1996). *Good Natured: The Origins of Right and Wrong in Humans and Other Animals* (p. 237). Harvard University Press.

36 *De Waal tells the story* De Waal (1996), p. 110.

37 *By age three, children* Liberman, Z., & Shaw, A. (2018). Secret to friendship: Children make inferences about friendship based on secret sharing. *Developmental Psychology, 54,* 2139–2151; Corson, K., & Colwell, M. J. (2013). Whispers in the ear: Preschool children's con-

ceptualisation of secrets and confidants. *Early Child Development and Care, 183,* 1215–1228.

37 *Young children do show* Warneken, F., & Tomasello, M. (2006). Altruistic helping in human infants and young chimpanzees. *Science, 311,* 1301–1303.

37 *and they seem surprised* Onishi & Baillargeon (2005).

37 *But if you tell a child* Wimmer, H., & Perner, J. (1983). Beliefs about beliefs: Representation and constraining function of wrong beliefs in young children's understanding of deception. *Cognition, 13,* 103–128.

38 *In a simplified false belief test* Gopnik, A., & Astington, J. W. (1988). Children's understanding of representational change and its relation to the understanding of false belief and the appearance-reality distinction. *Child Development, 59,* 26–37.

38 *At age three, however* Perner, J., Leekam, S. R., & Wimmer, H. (1987). Three-year-olds' difficulty with false belief: The case for a conceptual deficit. *British Journal of Developmental Psychology, 5,* 125–137.

38 *In another version of the study* Atance, C. M., & O'Neill, D. K. (2004). Acting and planning on the basis of a false belief: Its effects on 3-year-old children's reasoning about their own false beliefs. *Developmental Psychology, 40,* 953–964.

38 *In a follow-up study* Atance & O'Neill (2004).

39 *As children become better* Perner, J., Kloo, D., & Gornik, E. (2007). Episodic memory development: Theory of mind is part of re-experiencing experienced events. *Infant and Child Development: An International Journal of Research and Practice, 16,* 471–490.

39 *Children, at a young age* Flavell, J. H., Green, F. L., & Flavell, E. R. (1993). Children's understanding of the stream of consciousness. *Child Development, 64,* 387–398.

39 *"I'm going to ask you"* Flavell, J. H., Green, F. L., & Flavell, E. R. (2000). Development of children's awareness of their own thoughts. *Journal of Cognition and Development, 1,* 97–112.

40 *In a final version of the crayon* Atance & O'Neill (2004).

40 *The more children attend* Barresi, J. (2001). Extending self-consciousness into the future. In C. Moore & K. Lemmon (eds.), *The Self in Time: Developmental Perspectives* (pp. 141–162). Erlbaum.

41 *By around age six* Barresi (2001).

43   *children by age four* Wimmer & Perner (1983).

43   *When parents use more* Symons, D. K. (2004). Mental state discourse, theory of mind, and the internalization of self–other understanding. *Developmental Review, 24,* 159–188.

43   *Children with more siblings* Symons (2004).

44   *Friendships become forged* Furman, W., & Bierman, K. L. (1984). Children's conceptions of friendship: A multimethod study of developmental changes. *Developmental Psychology, 20,* 925–931.

44   *As children learn* Liberman & Shaw (2018).

44   *One young child described* Corson & Colwell (2013).

46   *As children pay more attention* Reese, E., Jack, F., & White, N. (2010). Origins of adolescents' autobiographical memories. *Cognitive Development, 25,* 352–367.

46   *Younger children have* Willoughby, K. A., Desrocher, M., Levine, B., & Rovet, J. F. (2012). Episodic and semantic autobiographical memory and everyday memory during late childhood and early adolescence. *Frontiers in Psychology, 3,* 53.

46   *Teenagers, in contrast* Chen, Y., McAnally, H. M., & Reese, E. (2013). Development in the organization of episodic memories in middle childhood and adolescence. *Frontiers in Behavioral Neuroscience, 7,* 84.

46   *Constructing and sharing* McLean, K. C. (2008). The emergence of narrative identity. *Social and Personality Psychology Compass, 2,* 1685–1702.

46   *and this remains true* McLean, K. C. (2005). Late adolescent identity development: Narrative meaning making and memory telling. *Developmental Psychology, 41,* 683–691.

47   *Teens can avoid* Darling, N., Cumsille, P., Caldwell, L. L., & Dowdy, B. (2006). Predictors of adolescents' disclosure to parents and perceived parental knowledge: Between- and within-person differences. *Journal of Youth and Adolescence, 35,* 659–670.

47   *And whereas younger children* Daddis, C., & Randolph, D. (2010). Dating and disclosure: Adolescent management of information regarding romantic involvement. *Journal of Adolescence, 33,* 309–320.

47   *Teenagers draw a line* Smetana, J. G. (1988). Adolescents' and parents' conceptions of parental authority. *Child Development, 59,* 321–335; Fuligni, A. J. (1998). Authority, autonomy, and parent-adolescent

conflict and cohesion: A study of adolescents from Mexican, Chinese, Filipino, and European backgrounds. *Developmental Psychology, 34*, 782–792.

48 *But if ongoing problems* Keijsers, L., Branje, S. J., VanderValk, I. E., & Meeus, W. (2010). Reciprocal effects between parental solicitation, parental control, adolescent disclosure, and adolescent delinquency. *Journal of Research on Adolescence, 20*, 88–113.

48 *Keeping certain secrets* Frijns, T., Finkenauer, C., Vermulst, A. A., & Engels, R. C. (2005). Keeping secrets from parents: Longitudinal associations of secrecy in adolescence. *Journal of Youth and Adolescence, 34*, 137–148.

49 *When teens expect* Smetana, J. G., Villalobos, M., Tasopoulos-Chan, M., Gettman, D. C., & Campione-Barr, N. (2009). Early and middle adolescents' disclosure to parents about activities in different domains. *Journal of Adolescence, 32*, 693–713.

49 *teens can construe* Kapetanovic, S., Bohlin, M., Skoog, T., & Gerdner, A. (2017). Structural relations between sources of parental knowledge, feelings of being overly controlled and risk behaviors in early adolescence. *Journal of Family Studies, 26*, 226–242; Hawk, S. T., Hale III, W. W., Raaijmakers, Q. A., & Meeus, W. (2008). Adolescents' perceptions of privacy invasion in reaction to parental solicitation and control. *Journal of Early Adolescence, 28*, 583–608.

49 *But when teens believe* Tilton-Weaver, L. (2014). Adolescents' information management: Comparing ideas about why adolescents disclose to or keep secrets from their parents. *Journal of Youth and Adolescence, 43*, 803–813.

50 *Secrecy is a common reaction* Wismeijer, A. A., Van Assen, M. A., & Bekker, M. H. (2014). The relations between secrecy, rejection sensitivity and autonomy-connectedness. *The Journal of General Psychology, 141*, 65–79; Cole, S. W., Kemeny, M. E., & Taylor, S. E. (1997). Social identity and physical health: Accelerated HIV progression in rejection-sensitive gay men. *Journal of Personality and Social Psychology, 72*, 320–335.

50 *secrecy is first clearly* Laird, R. D., Bridges, B. J., & Marsee, M. A. (2013). Secrets from friends and parents: Longitudinal links with depression and antisocial behavior. *Journal of Adolescence, 36*, 685–693.

CHAPTER 3: SECRETS ON THE MIND

51 *Dale Coventry and Jamie Kunz* 60 Minutes: "26-Year Secret Kept Innocent Man in Prison" (CBS television broadcast, March 9, 2008).

51 *Back in January 1982* Logan, A., with Falbaum, B. (2017). *Justice Failed: How "Legal Ethics" Kept Me in Prison for 26 Years.* Counterpoint; Winston, H. J. (2008). Learning from Alton Logan. *DePaul Journal for Social Justice, 2,* 173–189.

53 *Wilson, the real shooter* Conroy, J. (2007, Nov. 29). The persistence of Andrew Wilson. *The Chicago Reader.*

54 *The secret weighed heavily* 60 Minutes: "26-Year Secret Kept Innocent Man in Prison."

55 *In early November 1983* Pennebaker, J. W., & O'Heeron, R. C. (1984). Confiding in others and illness rate among spouses of suicide and accidental-death victims. *Journal of Abnormal Psychology, 93,* 473–476.

57 *talking about emotional burdens* Larson, D. G. (1985). Helper secrets: Invisible stressors in hospice work. *American Journal of Hospice Care, 2,* 35–40.

57 *the conclusions are clear* Larson, D. G., Chastain, R. L., Hoyt, W. T., & Ayzenberg, R. (2015). Self-concealment: Integrative review and working model. *Journal of Social and Clinical Psychology, 34,* 705–729.

58 *Kelly measured the extent* Kelly, A. E., & Yip, J. J. (2006). Is keeping a secret or being a secretive person linked to psychological symptoms? *Journal of Personality, 74,* 1349–1370.

60 *Just as we hoped* Slepian, M. L., Chun, J. S., & Mason, M. F. (2017). The experience of secrecy. *Journal of Personality and Social Psychology, 113,* 1–33.

62 *Studies estimate* Kane, M. J., Brown, L. H., McVay, J. C., Silvia, P. J., Myin-Germeys, I., & Kwapil, T. R. (2007). For whom the mind wanders, and when: An experience-sampling study of working memory and executive control in daily life. *Psychological Science, 18,* 614–621; Killingsworth, M. A., & Gilbert, D. T. (2010). A wandering mind is an unhappy mind. *Science, 330,* 932.

63 *a study that he conducted* Klinger, E. (1990). *Daydreaming.* Tarcher; Klinger, E. (1978). Modes of normal conscious flow. In K. S. Pope & J. L. Singer (eds.), *The Stream of Consciousness: Scientific Investigations into the Flow of Human Experience* (pp. 225–258). Plenum.

65 *Intentions, current goals* Klinger, E. (2013). Goal commitments and the content of thoughts and dreams: Basic principles. *Frontiers in Psychology, 4,* 415.

65 *This is what it means* Mason, M. F., & Reinholtz, N. (2015). Avenues down which a self-reminding mind can wander. *Motivation Science, 1,* 1–21.

66 *But this increased sensitivity* Slepian, M. L. (2021). A process model of having and keeping secrets. *Psychological Review.*

66 *When we are in a bad mood* Mayer, J. D., McCormick, L. J., & Strong, S. E. (1995). Mood-congruent memory and natural mood: New evidence. *Personality and Social Psychology Bulletin, 21, 736–746.*

66 *A natural response* Higgins, E. T., Klein, R., & Strauman, T. (1985). Self-concept discrepancy theory: A psychological model for distinguishing among different aspects of depression and anxiety. *Social Cognition, 3, 51–76.*

67 *Thinking about something over* Nolen-Hoeksema, S., Wisco, B. E., & Lyubomirsky, S. (2008). Rethinking rumination. *Perspectives on Psychological Science, 3, 400–424.*

68 *Try to not think about a white bear* Wegner, D. M., Schneider, D. J., Carter, S. R., & White, T. L. (1987). Paradoxical effects of thought suppression. *Journal of Personality and Social Psychology, 53, 5–13.*

69 *But when Anita Kelly asked* Kelly, A. E., & Kahn, J. H. (1994). Effects of suppression of personal intrusive thoughts. *Journal of Personality and Social Psychology, 66, 998–1006.*

69 *The more practice you have* Hu, X., Bergström, Z. M., Gagnepain, P., & Anderson, M. C. (2017). Suppressing unwanted memories reduces their unintended influences. *Current Directions in Psychological Science, 26, 197–206.*

69 *In a study of 800* Slepian, M. L., Greenaway, K. H., & Masicampo, E. J. (2020). Thinking through secrets: Rethinking the role of thought suppression in secrecy. *Personality and Social Psychology Bulletin, 46,* 1411–1427.

70 *my studies show* Slepian et al. (2020).

70 *But with some degree* Slepian et al. (2020).

70 *The compassionate response* Slepian, M. L., & Kirby, J. N. (2018). To

whom do we confide our secrets? *Personality and Social Psychology Bulletin, 44,* 1008–1023.

71 *If parents fail* Watkins, E. R., & Roberts, H. (2020). Reflecting on rumination: Consequences, causes, mechanisms and treatment of rumination. *Behaviour Research and Therapy, 127,* 103573.

71 *Rather than seek* Spasojević, J., & Alloy, L. B. (2002). Who becomes a depressive ruminator? Developmental antecedents of ruminative response style. *Journal of Cognitive Psychotherapy, 16,* 405–419.

71 *Rumination contributes* Watkins & Roberts (2020).

71 *people who tend to keep* Larson et al. (2015).

72 *Rather than look backward* Slepian et al. (2020).

CHAPTER 4: THE THREE DIMENSIONS OF SECRETS

76 *She asked participants* Strohminger, N., & Nichols, S. (2014). The essential moral self. *Cognition, 131,* 159–171.

76 *most people believe* Strohminger, N., Knobe, J., & Newman, G. (2017). The true self: A psychological concept distinct from the self. *Perspectives on Psychological Science, 12,* 551–560.

77 *Instead of attributing* Taylor, S. E., & Koivumaki, J. H. (1976). The perception of self and others: Acquaintanceship, affect, and actor-observer differences. *Journal of Personality and Social Psychology, 33,* 403–408; Green, S. P. (2003). *Underlying Processes as to Why the Fundamental Attribution Error Is Reduced in Close Relationships* (master's thesis, Miami University). ProQuest Dissertations Publishing.

77 *If people could get* Newman, G. E., Bloom, P., & Knobe, J. (2014). Value judgments and the true self. *Personality and Social Psychology Bulletin, 40,* 203–216.

78 *Participants from the United States* De Freitas, J., Sarkissian, H., Newman, G. E., Grossmann, I., De Brigard, F., Luco, A., & Knobe, J. (2018). Consistent belief in a good true self in misanthropes and three interdependent cultures. *Cognitive Science, 42,* 134–160.

78 *If you feel that you have changed* O'Brien, E., & Kardas, M. (2016). The implicit meaning of (my) change. *Journal of Personality and Social Psychology, 111,* 882–894.

79 *the three primary dimensions* Slepian, M. L., & Koch, A. (2021). Iden-

tifying the dimensions of secrets to reduce their harms. *Journal of Personality and Social Psychology, 120,* 1431–1456.

88 *Dilemmas like these* Cushman, F., & Young, L. (2009). The psychology of dilemmas and the philosophy of morality. *Ethical Theory and Moral Practice, 12,* 9–24; Greene, J. D. (2013). *Moral Tribes: Emotion, Reason, and the Gap Between Us and Them.* Penguin.

88 *Folks tend to have trouble* Haidt, J., Koller, S. H., & Dias, M. G. (1993). Affect, culture, and morality, or is it wrong to eat your dog? *Journal of Personality and Social Psychology, 65,* 613–628. See also Gray, K., & Wegner, D. M. (2011). Morality takes two: Dyadic morality and mind perception. In M. Mikulincer & P. R. Shaver (eds.), *The Social Psychology of Morality: Exploring the Causes of Good and Evil.* APA Press.

88 *By the aughts* Haidt, J. (2008). Morality. *Perspectives on Psychological Science, 3,* 65–72.

89 *Psychologist Wilhelm Hofmann* Hofmann, W., Wisneski, D. C., Brandt, M. J., & Skitka, L. J. (2014). Morality in everyday life. *Science, 345,* 1340–1343.

90 *conducted a small study* Bastian, B., Jetten, J., & Fasoli, F. (2011). Cleansing the soul by hurting the flesh. *Psychological Science, 22,* 334–335. See also Inbar, Y., Pizarro, D. A., Gilovich, T., & Ariely, D. (2013). Moral masochism: On the connection between guilt and self-punishment. *Emotion, 13,* 14–18; Nelissen, R.M.A., & Zeelenberg, M. (2009). When guilt evokes self-punishment: Evidence for the existence of a Dobby Effect. *Emotion, 9,* 118–122.

91 *We asked one group* Slepian, M. L., & Bastian, B. (2017). Truth or punishment: Secrecy and punishing the self. *Personality and Social Psychology Bulletin, 43,* 1596–1611.

91 *Shame is a particularly* Kim, S., Thibodeau, R., & Jorgensen, R. S. (2011). Shame, guilt, and depressive symptoms: A meta-analytic review. *Psychological Bulletin, 137,* 68–96.

92 *The current best estimate* Fincham, F. D., & May, R. W. (2017). Infidelity in romantic relationships. *Current Opinion in Psychology, 13,* 70–74; Whisman, M. A., Gordon, K. C., & Chatav, Y. (2007). Predicting sexual infidelity in a population-based sample of married individuals. *Journal of Family Psychology, 21,* 320–324; Marín, R. A., Chris-

tensen, A., & Atkins, D. C. (2014). Infidelity and behavioral couple therapy: Relationship outcomes over 5 years following therapy. *Couple and Family Psychology: Research and Practice, 3,* 1–12.

92 *A 2013 Pew survey* Extramarital affairs. Pew Research Center, Washington, D.C. (2014, Jan. 14).

92 *A few variables* Treas, J., & Giesen, D. (2000). Sexual infidelity among married and cohabiting Americans. *Journal of Marriage and Family, 62,* 48–60.

93 *a seasonal cycle* Adamopoulou, E. (2013). New facts on infidelity. *Economics Letters, 121,* 458–462.

93 *There is an increased incidence* Fincham & May (2017); Treas & Giesen (2000).

93 *that gap is closing* Abzug, R. (2016). Extramarital affairs as occupational hazard: A structural, ethical (cultural) model of opportunity. *Sexualities, 19,* 25–45.

93 *Across a study of 160* Betzig, L. (1989). Causes of conjugal dissolution: A cross-cultural study. *Current Anthropology, 30,* 654–676.

93 *an often-cited reason* Kurdek, L. A. (1991). The dissolution of gay and lesbian couples. *Journal of Social and Personal Relationships, 8,* 265–278.

93 *which presumably extends* Fincham & May (2017).

93 *a 2002 study found* Harris, C. R. (2002). Sexual and romantic jealousy in heterosexual and homosexual adults. *Psychological Science, 13,* 7–12.

94 *And relationships in their earliest* Wegner, D. M., Lane, J. D., & Dimitri, S. (1994). The allure of secret relationships. *Journal of Personality and Social Psychology, 66,* 287–300.

94 *At the same time* Foster, C. A., & Campbell, W. K. (2005). The adversity of secret relationships. *Personal Relationships, 12,* 125–143; Lehmiller, J. J. (2009). Secret romantic relationships: Consequences for personal and relational well-being. *Personality and Social Psychology Bulletin, 35,* 1452–1466.

95 *Finances are a common source* Garbinsky, E. N., Gladstone, J. J., Nikolova, H., & Olson, J. G. (2020). Love, lies, and money: Financial infidelity in romantic relationships. *Journal of Consumer Research, 47,* 1–24.

95 *A 2017 TD Bank survey* TD Bank (2017, Jan. 2), 2017 love and money survey.

95   *and another research team found* Jeanfreau, M. M., Noguchi, K., Mong, M. D., & Stadthagen-Gonzalez, H. (2018). Financial infidelity in couple relationships. *Journal of Financial Therapy, 9*, 1–20.

95   *A large phone survey* Mecia, T. (2015, Jan. 21). Financial infidelity poll: 6% hid bank account from spouse or partner.

97   *The good news is* Slepian & Koch (2021).

100   *You can still feel bad* Schmader, T., & Lickel, B. (2006). The approach and avoidance function of personal and vicarious shame and guilt. *Motivation and Emotion, 30*, 43–56.

103   *"I carefully evaluated"* Greenwald, G., MacAskill, E., & Poitras, L. (2013, June 10). Edward Snowden: The whistleblower behind the NSA surveillance revelations. *The Guardian.*

103   *"I'd draft manifestos"* Snowden, E. (2019). *Permanent Record* (p. 253). Pan Macmillan.

CHAPTER 5: CONCEALING OUR SECRETS

105   *Melody Casson was sixty-seven* Morris, S. (2015, July 21). Grandmother spared jail after admitting killing baby son 52 years ago. *The Guardian.*

107   *I asked 600 participants* Slepian, M. L., Chun, J. S., & Mason, M. F. (2017). The experience of secrecy. *Journal of Personality and Social Psychology, 113*, 1–33.

108   *16,000 words a day* Mehl, M. R., Vazire, S., Ramírez-Esparza, N., Slatcher, R. B., & Pennebaker, J. W. (2007). Are women really more talkative than men? *Science, 317*, 82.

109   *Actively avoiding specific* Caughlin, J. P., & Golish, T. D. (2002). An analysis of the association between topic avoidance and dissatisfaction: Comparing perceptual and interpersonal explanations. *Communication Monographs, 69*, 275–295.

109   *One study asked college students* Caughlin, J. P., & Afifi, T. D. (2004). When is topic avoidance unsatisfying? Examining moderators of the association between avoidance and dissatisfaction. *Human Communication Research, 30*, 479–513.

110   *conversation avoidance can be* Palomares, N. A., & Derman, D. (2019). Topic avoidance, goal understanding, and relational perceptions: Experimental evidence. *Communication Research, 46*, 735–756.

110 *A study of newlyweds* Finkenauer, C., Kerkhof, P., Righetti, F., & Branje, S. (2009). Living together apart: Perceived concealment as a signal of exclusion in marital relationships. *Personality and Social Psychology Bulletin, 35,* 1410–1422.

110 *When we think a partner* Finkenauer et al. (2009).

110 *Avoiding difficult conversations* Uysal, A., Lin, H. L., & Bush, A. L. (2012). The reciprocal cycle of self-concealment and trust in romantic relationships. *European Journal of Social Psychology, 42,* 844–851.

110 *Couples are often quite happy* Baxter, L. A., & Wilmot, W. W. (1985). Taboo topics in close relationships. *Journal of Social and Personal Relationships, 2,* 253–269.

110 *"The past should stay in the past"* Anderson, M., Kunkel, A., & Dennis, M. R. (2011). "Let's (not) talk about that": Bridging the past sexual experiences taboo to build healthy romantic relationships. *Journal of Sex Research, 48,* 381–391.

110 *These issues are often* Anderson et al. (2011).

111 *Suspicions of secrecy* Caughlin & Golish (2002).

111 *When you believe* Uysal et al. (2012).

111 *both are likely to perceive* Cole, T. (2001). Lying to the one you love: The use of deception in romantic relationships. *Journal of Social and Personal Relationships, 18,* 107–129.

112 *Larger conversations don't work* Cooney, G., Mastroianni, A. M., Abi-Esber, N., & Brooks, A. W. (2020). The many minds problem: Disclosure in dyadic versus group conversation. *Current Opinion in Psychology, 31,* 22–27.

112 *you might be worried that such a response* John, L. K., Barasz, K., & Norton, M. I. (2016). Hiding personal information reveals the worst. *Proceedings of the National Academy of Sciences, 113,* 954–959.

113 *One especially effective* Bitterly, T. B., & Schweitzer, M. E. (2020). The economic and interpersonal consequences of deflecting direct questions. *Journal of Personality and Social Psychology, 118,* 945–990.

113 *"Stealing secrets"* Snowden, E. (2019). *Permanent Record* (p. 257). Pan Macmillan.

113 *an abrupt shift of topic* Sun, K. Q., & Slepian, M. L. (2020). The conversations we seek to avoid. *Organizational Behavior and Human Decision Processes, 160,* 87–105.

114 *Research suggests* Rogers, T., & Norton, M. I. (2011). The artful dodger: Answering the wrong question the right way. *Journal of Experimental Psychology: Applied, 17*, 139–147.

114 *But if for some reason* Donovan-Kicken, E., Guinn, T. D., Romo, L. K., & Ciceraro, L. D. (2013). Thanks for asking, but let's talk about something else: Reactions to topic-avoidance messages that feature different interaction goals. *Communication Research, 40*, 308–336.

115 *your best bet* Donovan-Kicken et al. (2013).

116 *In his 1963 book* Goffman, E. (1963). *Stigma: Notes on the management of spoiled identity*. Prentice Hall.

117 *everyday situations* Slepian, M. L., & Jacoby-Senghor, D. (2021). Identity threats in everyday life: Distinguishing belonging from inclusion. *Social Psychological and Personality Science, 12*, 392–406.

118 *It took fifty years* Critcher, C. R., & Ferguson, M. J. (2014). The cost of keeping it hidden: Decomposing concealment reveals what makes it depleting. *Journal of Experimental Psychology: General, 143*, 721–735.

120 *A study published in the mid-'90s* Cole, S. W., Kemeny, M. E., Taylor, S. E., Visscher, B. R., & Fahey, J. L. (1996). Accelerated course of human immunodeficiency virus infection in gay men who conceal their homosexual identity. *Psychosomatic Medicine, 58*, 219–231; Cole, S. W., Kemeny, M. E., Taylor, S. E., & Visscher, B. R. (1996). Elevated physical health risk among gay men who conceal their homosexual identity. *Health Psychology, 15*, 243–251.

120 *One study captured just this* Beals, K. P., Peplau, L. A., & Gable, S. L. (2009). Stigma management and well-being: The role of perceived social support, emotional processing, and suppression. *Personality and Social Psychology Bulletin, 35*, 867–879.

121 *The study suggested* Legate, N., Ryan, R. M., & Weinstein, N. (2012). Is coming out always a "good thing"? Exploring the relations of autonomy support, outness, and wellness for lesbian, gay, and bisexual individuals. *Social Psychological and Personality Science, 3*, 145–152.

122 *In the late 1990s* Smart, L., & Wegner, D. M. (1999). Covering up what can't be seen: Concealable stigma and mental control. *Journal of Personality and Social Psychology, 77*, 474–486.

122 *What if people tried* Newheiser, A. K., & Barreto, M. (2014). Hidden costs of hiding stigma: Ironic interpersonal consequences of con-

cealing a stigmatized identity in social interactions. *Journal of Experimental Social Psychology, 52,* 58–70.

123 *To test whether concealment* Goh, J. X., Kort, D. N., Thurston, A. M., Benson, L. R., & Kaiser, C. R. (2019). Does concealing a sexual minority identity prevent exposure to prejudice? *Social Psychological and Personality Science, 10,* 1056–1064 (see footnote 4).

CHAPTER 6: CONFESSING AND CONFIDING

133 *The richness of our memory* Mahr, J. B., & Csibra, G. (2018). Why do we remember? The communicative function of episodic memory. *Behavioral and Brain Sciences, 41,* Article e1.

133 *By one estimate* Mahr, J. B., & Csibra, G. (2020). Witnessing, remembering, and testifying: Why the past is special for human beings. *Perspectives on Psychological Science, 15,* 428–443.

133 *Dessalles argues that human* Dessalles, J. L. (2007). *Why We Talk: The Evolutionary Origins of Language.* Oxford University Press.

133 *Telling stories is how* Mahr & Csibra (2018).

135 *With a treasure trove* Tamir, D. I., & Thornton, M. A. (2018). Modeling the predictive social mind. *Trends in Cognitive Sciences, 22,* 201–212.

136 *A long history of research* Willems, Y. E., Finkenauer, C., & Kerkhof, P. (2020). The role of disclosure in relationships. *Current Opinion in Psychology, 31,* 33–37.

138 *our predictions often miss* Epley, N. (2014). *Mindwise: How We Understand What Others Think, Believe, Feel, and Want.* Knopf.

138 *sing "The Star-Spangled Banner"* Epley, N., Savitsky, K., & Gilovich, T. (2002). Empathy neglect: Reconciling the spotlight effect and the correspondence bias. *Journal of Personality and Social Psychology, 83,* 300–312.

138 *and to sing along as best they can* Chambers, J. R., Epley, N., Savitsky, K., & Windschitl, P. D. (2008). Knowing too much: Using private knowledge to predict how one is viewed by others. *Psychological Science, 19,* 542–548.

138 *people were more charitable* Hall, J. A., & Taylor, S. E. (1976). When love is blind: Maintaining idealized images of one's spouse. *Human Relations, 29,* 751–761; Fiedler, K., Semin, G. R., Finkenauer, C., & Berkel, I. (1995). Actor-observer bias in close relationships: The role

of self-knowledge and self-related language. *Personality and Social Psychology Bulletin, 21*, 525–538; Prager, I. G., & Cutler, B. L. (1990). Attributing traits to oneself and to others: The role of acquaintance level. *Personality and Social Psychology Bulletin, 16*, 309–319; Taylor, S. E., & Koivumaki, J. H. (1976). The perception of self and others: Acquaintanceship, affect, and actor-observer differences. *Journal of Personality and Social Psychology, 33*, 403–408; Green, S. P. (2003). *The Underlying Processes as to Why the Fundamental Attribution Error Is Reduced in Close Relationships* (master's thesis, Miami University). ProQuest Dissertations Publishing.

140   *"It was like a light switch"* Asmelash, L. (2020, Aug. 31). Wedding announcement goes viral after groom's ex publicly reveals he cheated on her when he met the bride. CNN.

142   *Levine finds that when revealing* Levine, E. E. (2021). Community standards of deception. *Journal of Experimental Psychology: General.*

142   *A. J. Jacobs* Apton, D. (2009, Sept. 23). "Do not lie": Man lives by Biblical rules for a year. ABC News; Jacobs, A. J. (2007). *The Year of Living Biblically: One Man's Humble Quest to Follow the Bible as Literally as Possible.* Simon & Schuster.

147   *These questions are a variant* Aron, A., Melinat, E., Aron, E. N., Vallone, R., & Bator, R. (1997). The experimental generation of interpersonal closeness: A procedure and some preliminary findings. *Personality and Social Psychology Bulletin, 23*, 363–377.

147   *One woman's experience* Len Catron, M. (2015, Jan. 11). To fall in love with anyone, do this. *New York Times,* Section ST, p. 6.

148   *The research backs up* Welker, K. M., Baker, L., Padilla, A., Holmes, H., Aron, A., & Slatcher, R. B. (2014). Effects of self-disclosure and responsiveness between couples on passionate love within couples. *Personal Relationships, 21*, 692–708.

150   *I've surveyed thousands* Slepian, M. L., & Kirby, J. N. (2018). To whom do we confide our secrets? *Personality and Social Psychology Bulletin, 44*, 1008–1023.

150   *In a study of 200* Slepian, M. L., & Greenaway, K. H. (2018). The benefits and burdens of keeping others' secrets. *Journal of Experimental Social Psychology, 78*, 220–232.

151   *people don't typically judge* Savitsky, K., Epley, N., & Gilovich, T.

(2001). Do others judge us as harshly as we think? Overestimating the impact of our failures, shortcomings, and mishaps. *Journal of Personality and Social Psychology, 81,* 44–56.

151 *Luckily, our research* Salerno, J. M., & Slepian, M. L. (2022). Morality, punishment, and revealing other people's secrets. *Journal of Personality and Social Psychology, 122(4),* 606–633.

151 *People bond over gossip* Dunbar, R. I. (2004). Gossip in evolutionary perspective. *Review of General Psychology, 8,* 100–110.

151 *It also conveys* Feinberg, M., Willer, R., Stellar, J., & Keltner, D. (2012). The virtues of gossip: Reputational information sharing as prosocial behavior. *Journal of Personality and Social Psychology, 102,* 1015–1030; Salerno & Slepian (2022).

151 *people are less likely to confide* Slepian & Kirby (2018).

152 *the more someone is morally* Salerno & Slepian (2022).

152 *Despite how often* Slepian, M. L., & Moulton-Tetlock, E. (2019). Confiding secrets and well-being. *Social Psychological and Personality Science, 10,* 472–484.

152 *having a conversation about your secret* Slepian & Moulton-Tetlock (2019).

153 *Instead of talking* Pennebaker, J. W., & Beall, S. K. (1986). Confronting a traumatic event: Toward an understanding of inhibition and disease. *Journal of Abnormal Psychology, 95,* 274–281.

153 *Pennebaker's follow-up studies* Pennebaker, J. W., Mayne, T. J., & Francis, M. E. (1997). Linguistic predictors of adaptive bereavement. *Journal of Personality and Social Psychology, 72,* 863–871.

153 *the journal just becomes* Ullrich, P. M., & Lutgendorf, S. K. (2002). Journaling about stressful events: Effects of cognitive processing and emotional expression. *Annals of Behavioral Medicine, 24,* 244–250.

153 *Even more beneficial* Pennebaker, J. W. (1993). Putting stress into words: Health, linguistic, and therapeutic implications. *Behaviour Research and Therapy, 31,* 539–548; Pennebaker, J. W., & Francis, M. E. (1996). Cognitive, emotional, and language processes in disclosure. *Cognition and Emotion, 10,* 601–626.

154 *Pennebaker is the first to admit* Smyth, J. M., & Pennebaker, J. W. (2008). Exploring the boundary conditions of expressive writing: In search of the right recipe. *British Journal of Health Psychology, 13,* 1–7.

(Although, based on the author order here, you might argue that Pennebaker is the second to admit that journaling doesn't work for everyone.)

154 *writing about positive life events* Burton, C. M., & King, L. A. (2004). The health benefits of writing about intensely positive experiences. *Journal of Research in Personality, 38,* 150–163.

154 *And writing about coping* Greenberg, M. A., Wortman, C. B., & Stone, A. A. (1996). Emotional expression and physical health: Revising traumatic memories or fostering self-regulation? *Journal of Personality and Social Psychology, 71,* 588–602.

154 *What is helpful about* Hemenover, S. H. (2003). The good, the bad, and the healthy: Impacts of emotional disclosure of trauma on resilient self-concept and psychological distress. *Personality and Social Psychology Bulletin, 29,* 1236–1244; King, L. A. (2001). The health benefits of writing about life goals. *Personality and Social Psychology Bulletin, 27,* 798–807; Creswell, J. D., Lam, S., Stanton, A. L., Taylor, S. E., Bower, J. E., & Sherman, D. K. (2007). Does self-affirmation, cognitive processing, or discovery of meaning explain cancer-related health benefits of expressive writing? *Personality and Social Psychology Bulletin, 33,* 238–250; Bonanno, G. A. (2004). Loss, trauma, and human resilience: Have we underestimated the human capacity to thrive after extremely aversive events? *American Psychologist, 59,* 20–28.

154 *If putting words down* Pennebaker, J. W., & Smyth, J. M. (2016). *Opening Up by Writing It Down: How Expressive Writing Improves Health and Eases Emotional Pain.* Guilford Publications.

154 *sending an anonymous postcard* Warren, F. (2005). PostSecret: Extraordinary confessions from ordinary lives. William Morrow.

155 *revealing a secret anonymously* Slepian, M. L., Masicampo, E. J., & Ambady, N. (2014). Relieving the burdens of secrecy: Revealing secrets influences judgments of hill slant and distance. *Social Psychological and Personality Science, 5,* 293–300.

155 *Without another person* Nils, F., & Rimé, B. (2012). Beyond the myth of venting: Social sharing modes determine the benefits of emotional disclosure. *European Journal of Social Psychology, 42,* 672–681.

155 *Another person can challenge* Lepore, S. J., Fernandez–Berrocal, P., Ragan, J., & Ramos, N. (2004). It's not that bad: Social challenges to

emotional disclosure enhance adjustment to stress. *Anxiety, Stress & Coping, 17,* 341–361.

155 *practical support and fresh perspectives* Slepian & Moulton-Tetlock (2019).

CHAPTER 7: POSITIVE SECRETS

158 *It doesn't have to be* Langston, C. A. (1994). Capitalizing on and coping with daily-life events: Expressive responses to positive events. *Journal of Personality and Social Psychology, 67,* 1112–1125; Gable, S. L., Reis, H. T., Impett, E. A., & Asher, E. R. (2004). What do you do when things go right? The intrapersonal and interpersonal benefits of sharing positive events. *Journal of Personality and Social Psychology, 87,* 228–245; Gable, S. L., & Reis, H. T. (2010). Good news! Capitalizing on positive events in an interpersonal context. In M. P. Zanna (ed.), *Advances in Experimental Social Psychology* (vol. 42, pp. 195–257). Academic Press.

159 *who would you visit first?* Loewenstein, G. F., & Prelec, D. (1993). Preferences for sequences of outcomes. *Psychological Review, 100,* 91–108.

160 *a five-minute speech* Monfort, S. S., Stroup, H. E., & Waugh, C. E. (2015). The impact of anticipating positive events on responses to stress. *Journal of Experimental Social Psychology, 58,* 11–22.

161 *lottery ticket* Kocher, M. G., Krawczyk, M., & van Winden, F. (2014). "Let me dream on!" Anticipatory emotions and preference for timing in lotteries. *Journal of Economic Behavior & Organization, 98,* 29–40.

161 *uncertainty magnifies* Gilbert, D. (2006). *Stumbling on Happiness.* Vintage.

161 *We know this from a study* Kurtz, J. L., Wilson, T. D., & Gilbert, D. T. (2007). Quantity versus uncertainty: When winning one prize is better than winning two. *Journal of Experimental Social Psychology, 43,* 979–985.

162 **Savoring** Bryant, F. B., & Veroff, J. (2017). *Savoring: A New Model of Positive Experience.* Psychology Press.

163 *the tendency to savor the good* Hurley, D. B., & Kwon, P. (2013). Savoring helps most when you have little: Interaction between savoring the moment and uplifts on positive affect and satisfaction with life. *Journal of Happiness Studies, 14,* 1261–1271; Jose, P. E., Lim, B. T., &

Bryant, F. B. (2012). Does savoring increase happiness? A daily diary study. *The Journal of Positive Psychology, 7,* 176–187.

163 *the words people used* Augustine, A. A., Mehl, M. R., & Larsen, R. J. (2011). A positivity bias in written and spoken English and its moderation by personality and gender. *Social Psychological and Personality Science, 2,* 508–515.

163 *For example, "good"* Rozin, P., Berman, L., & Royzman, E. (2010). Biases in use of positive and negative words across twenty natural languages. *Cognition and Emotion, 24,* 536–548.

163 *Another study buzzed* Brans, K., Koval, P., Verduyn, P., Lim, Y. L., & Kuppens, P. (2013). The regulation of negative and positive affect in daily life. *Emotion, 13,* 926–939.

164 *negative events feel more newsworthy* Koch, A., Alves, H., Krüger, T., & Unkelbach, C. (2016). A general valence asymmetry in similarity: Good is more alike than bad. *Journal of Experimental Psychology: Learning, Memory, and Cognition, 42,* 1171–1192.

164 *In 2013, two avid* Pickens, P. (2018, April 4). Couple finishes five-year NHL arena tour with engagement in Calgary. NHL.com.

164 *an entire industry of proposals* de la Cretaz, B. (2019, March 31). Thinking of a jumbotron proposal? Some say, ugh. Others, say yes. *New York Times,* Section ST, p. 15.

164 *"It's not about the public"* Pickens (2018, April 4).

165 *The emotional experience* Verduyn, P., & Lavrijsen, S. (2015). Which emotions last longest and why: The role of event importance and rumination. *Motivation and Emotion, 39,* 119–127.

165 *Surprises, by definition* Loewenstein, J. (2019). Surprise, recipes for surprise, and social influence. *Topics in Cognitive Science, 11,* 178–193.

165 *a TV commercial* Davila, F. (2007, Feb. 7). Very public proposal will be something to "tell the kids about." *The Seattle Times.*

165 *People have been wrapping* Colwell, C. (2017, Dec. 19). Why do we wrap gifts? *Sapiens.*

167 *Not discussing marriage* Hoplock, L. (2016). *Will She Say Yes? A Content Analysis of Accepted and Rejected Marriage Proposals* (doctoral dissertation, University of Victoria).

169 *we feel more capable* Schwarzer, R. (ed.). (2014). *Self-efficacy: Thought Control of Action.* Taylor & Francis; Chwalisz, K., Altmaier, E. M., &

Russell, D. W. (1992). Causal attributions, self-efficacy cognitions, and coping with stress. *Journal of Social and Clinical Psychology, 11,* 377–400; Taylor, S. E., & Armor, D. A. (1996). Positive illusions and coping with adversity. *Journal of Personality, 64,* 873–898.

169 *happier and healthier* Mendes de Leon, C. F., Seeman, T. E., Baker, D. I., Richardson, E. D., & Tinetti, M. E. (1996). Self-efficacy, physical decline, and change in functioning in community-living elders: A prospective study. *The Journals of Gerontology Series B: Psychological Sciences and Social Sciences, 51,* S183–S190.

169 *They also live longer* Krause, N., & Shaw, B. A. (2000). Role-specific feelings of control and mortality. *Psychology and Aging, 15,* 617–626.

172 *When someone responds negatively* Gable et al. (2004).

CHAPTER 8: CULTURE AND COPING

173 *Lulu Wang was born* Glass, I. (producer). (2016, April 22). 585: "In Defense of Ignorance." *This American Life.*

177 *relational mobility* Kito, M., Yuki, M., & Thomson, R. (2017). Relational mobility and close relationships: A socioecological approach to explain cross-cultural differences. *Personal Relationships, 24,* 114–130.

178 *In low relational mobility environments* Thomson, R., Yuki, M., Talhelm, T., Schug, J., Kito, M., Ayanian, A., . . . Visserman, M. L. (2018). Relational mobility predicts social behaviors in 39 countries and is tied to historical farming and threat. *Proceedings of the National Academy of Sciences, 115,* 7521–7526.

178 *these tight-knit networks* Kito et al. (2017).

178 *personal choices and traveling* Schug, J., Yuki, M., Horikawa, H., & Takemura, K. (2009). Similarity attraction and actually selecting similar others: How cross-societal differences in relational mobility affect interpersonal similarity in Japan and the United States. *Asian Journal of Social Psychology, 12,* 95–103.

178 *more trusting of strangers* Thomson et al. (2018).

178 *each country contains* Thomson et al. (2018); Morris, M. W., Chiu, C. Y., & Liu, Z. (2015). Polycultural psychology. *Annual Review of Psychology, 66,* 631–659.

179 *more social risks* Li, L.M.W., Hamamura, T., & Adams, G. (2016). Re-

lational mobility increases social (but not other) risk propensity. *Journal of Behavioral Decision Making, 29,* 481–488.

179 *In high relational mobility environments* Yamada, J., Kito, M., & Yuki, M. (2017). Passion, relational mobility, and proof of commitment: A comparative socio–ecological analysis of an adaptive emotion in a sexual market. *Evolutionary Psychology, 15,* 1–8; Li, L.M.W., Adams, G., Kurtiş, T., & Hamamura, T. (2015). Beware of friends: The cultural psychology of relational mobility and cautious intimacy. *Asian Journal of Social Psychology, 18,* 124–133; Schug, J., Yuki, M., & Maddux, W. W. (2010). Relational mobility explains between- and within-culture differences in self-disclosure toward close friends. *Psychological Science, 21,* 1471–1478.

181 *"If I talked to my American friends"* Dessem, M. (2020, Jan. 5). Lulu Wang's grandmother learned she had cancer from *The Farewell. Slate.*

181 *"You want to tell Nai Nai"* Wang, L. (director). (2019). *The Farewell* (film).

181 *hallmarks of collectivism* Brewer, M. B., & Chen, Y. R. (2007). Where (who) are collectives in collectivism? Toward conceptual clarification of individualism and collectivism. *Psychological Review, 114,* 133–151.

182 *The more the participants were religious* Cukur, C. S., De Guzman, M.R.T., & Carlo, G. (2004). Religiosity, values, and horizontal and vertical individualism—Collectivism: A study of Turkey, the United States, and the Philippines. *The Journal of Social Psychology, 144,* 613–634.

182 *Folks who agree* Kashima, Y., Yamaguchi, S., Kim, U., Choi, S. C., Gelfand, M. J., & Yuki, M. (1995). Culture, gender, and self: A perspective from individualism-collectivism research. *Journal of Personality and Social Psychology, 69,* 925–937.

183 *they felt more authentic* McDonald, R. I., Salerno, J. M., Greenaway, K. H., & Slepian, M. L. (2020). Motivated secrecy: Politics, relationships, and regrets. *Motivation Science, 6,* 61–78.

185 *The more you are concerned* Kim, H. S., & Chu, T. Q. (2011). Cultural variation in the motivation of self-expression. In D. A. Dunning (ed.), *Frontiers of Social Psychology: Social Motivation* (pp. 57–77). Psychology Press.

185 *At one extreme* Kim, H. S., & Markus, H. R. (2002). Freedom of speech and freedom of silence: An analysis of talking as a cultural practice. In R. Shweder, M. Minow, & H. R. Markus (eds.), *Engaging Cultural Differences: The Multicultural Challenge in Liberal Democracies* (pp. 432–452). Russell Sage Foundation.

185 *those in collectivistic cultures* LeClair, J., Janusonis, S., & Kim, H. S. (2014). Gene–culture interactions: A multi-gene approach. *Culture and Brain, 2,* 122–140; Kim, H. S., & Ko, D. (2007). Culture and self-expression. In C. Sedikides & S. J. Spencer (eds.), *Frontiers of Social Psychology: The Self* (pp. 325–342). Psychology Press.

186 *These relational concerns* Kim, H. S., Sherman, D. K., Ko, D., & Taylor, S. E. (2006). Pursuit of comfort and pursuit of harmony: Culture, relationships, and social support seeking. *Personality and Social Psychology Bulletin, 32,* 1595–1607; Kim, H. S., Sherman, D. K., & Taylor, S. E. (2008). Culture and social support. *American Psychologist, 63,* 518–526.

186 *She finds that in Asian* Kim et al. (2008).

186 *In cultures that prioritize* Taylor, S. E., Sherman, D. K., Kim, H. S., Jarcho, J., Takagi, K., & Dunagan, M. S. (2004). Culture and social support: Who seeks it and why? *Journal of Personality and Social Psychology, 87,* 354–362.

187 *Kim's study found* Kim et al. (2008); Taylor, S. E., Welch, W. T., Kim, H. S., & Sherman, D. K. (2007). Cultural differences in the impact of social support on psychological and biological stress responses. *Psychological Science, 18,* 831–837; Chen, J. M., Kim, H. S., Sherman, D. K., & Hashimoto, T. (2015). Cultural differences in support provision: The importance of relationship quality. *Personality and Social Psychology Bulletin, 41,* 1575–1589.

187 *in open-ended responses* Taylor et al. (2004).

187 *Another study asked the extent* Taylor et al. (2004).

188 *While the nature of support-seeking* Campos, B., & Kim, H. S. (2017). Incorporating the cultural diversity of family and close relationships into the study of health. *American Psychologist, 72,* 543–554.

LIST OF ILLUSTRATIONS

pages 16–17   The Secrets People Commonly Keep

page 81   Subway Stations

page 82   A Map of Common Secrets

INDEX

NOTE: Page references in *italics* refer to figures.

38 common categories of secrets,
13–15, *16, 17,* 18–21, 24, 59–61,
107, 125, 148, 151, 175–176
3D map of secrets, 81–85

abilities (development of)
autobiographical memory, 46
cognitive, 27–32
communication, 37
concealment, 35–36, 41
perspective taking, 42
placing trust in others, 50
privacy, 47
self-consciousness, 42
self-expression, 46
storytelling, 43–44
suppressing thoughts, 69
understanding mental states, 40, 43
understanding other minds, 34
abortion, 13, *16, 17,* 19
abuse, 70
accidents, 35–36
addiction, 13, *16, 17, 82,* 86, 107
admissions, 49, 70, 138–139, 150–152

adolescence. *See* teenagers
adultery. *See* infidelity
advice, 50, 131, 145, 152, 162–163
affairs. *See* infidelity
agreeableness (personality trait), 24–25
air metaphor for culture, 175, *176*
ambitions (plans, goals), 14, 15, *16, 17,*
18, 42, *82,* 108
anticipation
of positive events, 159–160
of positive secrets, 165–166, 168, 170
of unknown outcomes, 161–162
anxiety, 7, 65, 71
Aron, Arthur, 147
The Art of Choosing (Iyengar), 137
Ashley Madison (website), 92
assertiveness, 150
attention, 27, 40, 62–65, 113, 120, 162,
164
Aufort, Adrien, 59–61
authentic self, 116–117, 182–183. *See
also* self; true self
autonomy, 47, 103, 172
awkward, feeling, 7, 45, 112, 124, 181

babies, cognitive development, 32–35

Bastian, Brock, 90, 91

behavior, disposition *vs.* situation, 77–78, 140

beliefs, concealing, 14, *16*, *17*, *82*, 86, 94, 108

BoJack Horseman (television show), 67–68, 70

burden of secrets
  frequency of thinking about secret, 12, 45–46, 54
  hill slant studies about, 8–12, 61
  intention and, 72–73
  from isolation, 126–128
  on our own time, 104
  seeking resolution for, 125–126
  unburdening by sharing, 189–194
  well-being and, 4, 8–12, 25, 55–59, 87–96, 120, 152–156

California, ix, 78, 100–101, 117–118, 166

career goals. *See* profession/goals

Carter, Jimmy, 18

Casson, Melody, 105–106, 128

Catholic confession, 91

Catron, Mandy Len, 147–148

Chan, Joanna, 164–165

character, 75–78. *See also* self

Chavez, Matthew, 129–131

cheating on romantic partners. *See* infidelity; romantic relationships and sex

cheating (work or school), 13, *16*, *17*, 18, *82*, 86, 87, 95. *See also* profession/goals

children and child development, 26–50
  babies and, 32–35
  age fifteen to eighteen months, 33–34
  age two, 35
  age three to five, 35–41
  age six to eight, 41–42

cognitive ability of chimpanzees *vs.* humans, 27–32, 33, 34, 36–37, 42–44
  communicating difficult thoughts/ experiences, 70–71
  concealment by children, 35–36
  false belief test and, 30–35, 37–39, 40, 43
  inference and, 26–27
  middle childhood and, 41–44
  positive social power of secrets and, 192–193
  pre-school children and, 35–41
  teenagers and, 45–50

Chimpanzee Politics (de Waal), 29

chimpanzees, cognitive ability of, 27–32, 33, 34, 36–37, 42–44

China, culture and secrecy, 173–176, 181, 183–184

cognition
  cognitive development (*See* children and child development)
  intention and cognitive ability, 27–32
  journaling for cognitive processing, 153–154, 156
  memory, 39–41, 46, 131–136
  mind-wandering tendency, 62–66
  neuroimaging studies of imagining sensation, 8–9
  rumination and, 66–72

cohabiting, 92, 95

collectivism, 181–186

Colombia, culture and core selves as moral, 78

Columbia University, ix, 59, 63, 116, 133, 137

Colwell, Chip, 165

communication. *See also* concealment; confession; confiding; disclosure
  about difficult thoughts/ experiences, 70–71

avoiding conversation, 108, 109–111
childhood development, 37, 43, 46–50
desire to share good news, 157–159 (*See also* positive secrets)
difficult conversations, 110–111
dodging questions and, 112–115
sharing secrets, 44
speaking about grief, 55–59
storytelling, 43–44, 46, 131–134
compassion, 49, 70, 139, 150
"complete secrets," 19–20, 108
concealment, 105–128
  altering language, 120
  avoiding conversation topics, 108, 109–111
  being asked a question, 108
  of children's accidents, 35–36
  college major study, 122–123
  vs. confessing secrets, 105–106, 124–126, 128
  dodging questions, 112–115
  due to shame/embarrassment, 57–59
  harm of, 126–128
  hiding the evidence, 107–108
  vs. mind-wandering, 127–128
  secrets most commonly concealed, 108–109
  self-monitoring and, 115–121
  sexual orientation, 120–123
  success of, 121–124
concerns, current, 64–66
confession, 140–146
  accidental revelation, 143–144
  consequences of, 141, 145
  deciding to confess, 141–146
  dilemmas, 141–142, 145–146
  preparing someone for, 144–145
  timing of, 144
confidants, 150–152, 156

"confided secrets"
  by children, 37, 43–44
  statistics of, 19–20
confiding, 146–152
  being confided in, 151
  the best time to confide, 150
  consequences of, 152
  feeling better, 149
  reciprocal nature of, 149
  in a stranger (study), 147–149
  waiting to confide, 146–147
  whom to confide in, 150–152
conflict avoidance
  financial secrets and, 95
  secrecy *vs.* privacy within relationships, 22–23, 109–111
conscientiousness (personality trait), 24–25
control, feeling in, 47, 67, 97–98, 169–170
conversation, xii, 4, 7–8, 23, 45, 49, 66, 108–128, 134–136, 144–145, 155, 163, 177. *See also* communication; concealment
  number of words spoken per day, 108
  positivity of words spoken, 163
coping, 71, 91–92, 97–104, 152–154, 171
  confiding, 152
  journaling for cognitive processing, 153–154, 156
  speaking about grief, 55–59, 152
  unhealthy style of, 58–59
coping compass
  choosing path forward with, 104
  defined, 98
  insight, understanding (strategy 3), 98, 102–104
  harm (strategy 1), 97–100
  protection of others (strategy 2), 98, 100–102

coping compass (*cont.*)
for shame, isolation, lack of insight,
96–97
Cornell University, 118
counternormative behavior, 14, *16,*
17, 82
courage, 111, 140, 179
Coventry, Dale, 51–55
covering, 118
crayons/candles studies, 37–39, 40–41
crime, and confession of, 3, 51–52,
105–106, 128
Critcher, Clayton, 118–119
criteria for harm of secret. *See* three
dimensions of secrets
culture, 173–188
air metaphor for, 175, 176
collectivism, 181–186
common types of secrets and, 17
emotional expression and, 184–188
Log kya kahenge ("What will people
think?"), 137–140
perception of morality, 78–80,
92–93
relational mobility, 177–180
secrets about terminal illness,
173–176, 181, 183–184
sexual infidelity and, 92, 93
values, 180–184

decisiveness, 150
defining secrecy, 7–8
depression, 48, 67, 71, 73
Dessalles, Jean-Louis, 133
de Waal, Franz, 29, 36–37
dimensions of secrets. *See* three
dimensions of secrets
disclosure, 129–156
in adolescence, 49–50
anonymous methods of, 129–131,
152–156
barriers to, 139–149
becoming known, 135–136

vs. concealing, 105–106, 124–126,
128
culture and managing emotion,
184–188
learn about ourselves, 135–136
learn from each other, 133–134
perception by others about, 136–
140
relationship health, 49–50, 136
self-disclosure and culture, 179–180
self-knowledge and, 134–136
sharing past experiences and,
131–134
social connection, 136, 149
discontent. *See* physical discontent;
profession/work/school
discontent; romantic discontent;
social discontent
distraction, 62–66
diversity, 60, 116–117, 178
divorce, 93, 178
DNA testing, 143
donor conception example, ix–xiii,
11, 72–73, 124–125, 143–144,
180–181, 190–194
drug use, 13, 15, *16, 17, 82*

eating disorder study, 122
embarrassment, 15, 42, 57–58, 115
Emory University, 29
emotion and feeling
culture and expression of, 184–188
mood, 66–67, 163–164
positive secrets, 164–170
three dimensions of secrets, 96
emotional harm, 13, *16, 17, 82,* 86
emotional infidelity, 13, 15, *16, 17,*
18–19, *82,* 86, 93
emotional stability (personality
trait), 24
employment, hidden, 14, *16, 17, 82,*
86, 95
Epley, Nicholas, 138

Escobar, Pablo, 107
evidence, hiding. *See* concealment
extra-relational thoughts, 13, *16, 17,*
    18–20, *82*
extraversion (personality trait), 24–25

facial expression, 122, 184–185
false belief test, 30–35, 37–39, 40, 43
family secrets. *See also* culture; *The*
    Farewell
    common secrets about, 14, *16, 17,*
        18, *82,* 86
    donor conception example, ix–xiii,
        11, 72–73, 124–125, 143–144,
        180–181, 192–194
The Farewell (film), 174–175
Fast Friends Procedure, 147–148
    in couples, 149
    falling in love, 147–148
    and social closeness, 147
fatigue, 9, 119–120
fears, 7, 15, 23, 50, 58, 77, 111, 134, 138
Fight Club (film), 108
finances
    common secrets about, 14, *16, 17,*
        18, 22, *82*
    concealment about, 108–109
    goal-related dimension of secrets,
        85–87, 94–96
fireplace illusion, 8–9
France, morality of infidelity, 92
frequency of concealing, 126
frequency of thoughts, 51–55, 59–61,
    66–72
friendships, 44–50

gender
    concealment of partner preference,
        119–120
    identity, 14, *16, 17, 82,* 123
    sexual infidelity and, 93
genetic testing, 190–194
gifts as positive secrets, 164–170

Gilbert, Daniel, 161–162
goals, 64–66. *See also* ambitions,
    concerns; profession/goals
Goffman, Erving, 116, 118
Goh, Jin, 123
Good Natured (de Waal), 36–37
gossip, 151–152
The Graduate (film), 100–102
grief, coping with, 55–59, 152–156
guilt, 90, 100, 180

habit, 13, *16, 17, 82,* 94, 107
Haidt, Jonathan, 89
harm (coping strategy 1), 97–100
HBO, 3
height/jumping fear example, 64–65
helplessness, 67, 90–91
hiding objects, 27, 30, 32, 34, 42, 107
Higgins, Tory, 133–134
hill slant studies, 8–12, 61
hobbies
    common secrets about, 14, *16,*
        17, 82
    concealment about, 86, 94, 108,
        108–109, 118
    as secret joys, 171
Hofmann, Wilhelm, 89
honesty, expectations of, 102, 141–142
Hope, Edgar, 51–55
hospice workers, 56–57, 71
hostel example, relational mobility,
    177, 179
humor, to avoid questions, 112–114

illegal behavior, 13, 15, *16, 17,* 20,
    82, 86
imaginary pill study, 76
inauthentic, 118, 182–184, 188
India, culture and secrecy, 137–140
infidelity, 10, 13, *16, 17,* 18–19, 22, *82,*
    141, 145
    consequences of, 93, 141
    emotional *vs.* sexual, 15, 93

infidelity (*cont.*)
  gender, 93
  how common, 92
  morality of, culture, 92
  predictors of, 92–93
  relationship with someone
    cheating on a partner, 13, *16,*
    17, 82
  whether to confess, 145–146
inner goodness, 77
inner monologue, 67–68, 70
inner struggles, 57–58, 186
inner worlds, xii, 40–44, 133, 189,
  191–193
insight. *See also* coping compass
  lack of, 96–104
  understanding (coping strategy 3),
    98, 102–104
intention
  burden of secrets and, 72–73
  cognitive ability, 27–32 (*See
    also* children and child
    development)
  frequency of thoughts and, 51–55,
    59–61, 66–72 (*See also* secrets on
    the mind)
  heights and jumping example,
    64–65
  mind-wandering, 65–66
  secrecy *vs.* privacy, 8, 13, 19–23,
    109–111
  unfulfilled goals, 65
internet, anonymous confession via,
  155
intrusive thoughts. *See* rumination
isolation
  burden of secrets from, 126–128
  coping, 188
  coping compass for, 96–104
  relational mobility and, 180
  Snowden example, 6–7
  solitude *vs.*, 170–172
Italy, morality of infidelity, 92

"It's the End of the World as We Know
  It" (R.E.M.) study, 138
Iyengar, Sheena, 137, 139, 147

Jacobs, A. J., 142
Jacoby-Senghor, Drew, 117
job. *See* profession/goals
jokes. *See* humor
journaling, 153–154, 156
  caveats, 154–156
  grief study, 55–56
  how it helps, 154
  how to, 153–154
  trauma study, 153
jumbotron proposals, 164–165
jumping/height fear example, 64–65

keeping secrets for others, 48, 150
Kellogg School of Management,
  116–117
Kelly, Anita, 58, 69
Kim, Heejung, 186–188
Klinger, Eric, 63
Koch, Alex, 83
Korea, culture and secrecy, 186–188
Kunz, Jamie, 51–55, 61

lack of insight. *See* insight
language. *See* communication
Larson, Dale, 56–57, 71
Levine, Emma, 141–142
lie, lying (as a secret), 13, *16, 17,* 18, *82*
Logan, Alton, 51–55, 61
Log kya kahenge ("What will people
  think?"), 137–140
Love, Simon (film), 45, 50, 57
Lu, Hong, 173–176, 181, 183–184

marriage proposal
  common secrets about, 14, *16,*
    17, 19
  as positive secret, 164–170
Mason, Malia, 63

McDonald's shooting example,
  51–55, 61
memory, 131–136
  autobiographical memory, 46
  development of, 39–40, 46
  episodic memory, 132–133
  knowledge, 39, 41, 43, 46, 132–133
  richness of, 133
  semantic memory, 132
  sharing and storytelling, 133
  as special, 131–132
mental experiences, attention to,
  39–43, 46
mental health, concealing, 13, 15, 16,
  17, 82, 86, 94, 96, 108. See also
  well-being (mental and physical)
metaphor, 9
middle childhood, cognitive
  development, 41–44
mind-wandering, 62–66
Mindwise (Epley), 138
mischief, 35
mistakes, 63–64, 78, 98–100, 145
mock interview studies, 4, 118–119,
  122–123
"Modern Love" (Catron, New York Times
  essay), 147–148
money. See finances
morality. See also three dimensions of
  secrets
  character and, 75–78
  chicken story, 88
  confiding secrets and, 151–152
  development, 88
  as dimension of secrets, 85, 87–92
  everyday morality, 89
  moral dilemmas, 88, 141, 145, 175
  moral dumbfounding, 89
  moral outrage, 152
  perception of, 78–80, 92–93
  revealing secrets, 151–152
  secrecy vs. privacy, 22
  trolley problem, 87–88

Moreno, Nikyta, 140–141, 145
Morris, Julie, 164–165
multidimensional scaling, 83

"Nai Nai" (unaware of having cancer),
  173–176, 181, 183–184
National Security Agency (NSA), 4–6,
  7, 12, 102–104
negative thinking
  rumination, 66–72
  when alone with a secret, 140
neuroticism (personality trait), 24–25
New York City, ix, 59, 80, 85, 129
New York Times, 147–148
newlyweds, 110
NPR, 174, 184
number of secrets, 17, 20, 24–25, 180
number of thoughts, 63

OCEAN (Openness,
  Conscientiousness, Extraversion,
  Agreeableness, Neuroticism),
  24–25
Omertà, 4
openness (personality trait), 24

Pakistan, morality of infidelity, 92
parenting, 49, 70–71
Pennebaker, James, 55–56, 152–153
perception. See also culture; self; social
  support; trust
  imagining sensation, 8–12
  of morality, 78–80, 92–93
  by others, about concealment,
  121–124
  by others, about confession,
  136–140, 151
  of others' concealment, 110
  self-consciousness of children, 42
performance at work/school, poor, 14,
  16, 17, 82, 95
personal story, 14, 16, 17, 82, 86
personality traits, 24–25, 176

perspective-taking, 27, 42–44
perspectives, new, 152, 154, 156
Pew survey (2013), 92
Philippines, culture and collectivism, 182
Phillips, Katherine, 116–117
physical discontent, 13, *16*, *17*, *82*
physical harm, 13, *16*, *17*, *82*, 86
Plato, 96
"play a trick" study, 33–34
Playboy magazine, 18
political views, hidden, 14
positive secrets, 157–172
   anticipating pleasant experiences, 159–162, 165, 168, 170
   gifts, proposals, surprises as, 164–170
   savoring, 162–164
   secret joys as, 170–172
   sharing good news, 157–159
possessions, hidden. *See* hobbies
PostSecret project, 155
prediction of others' perceptions, 138–140
preference, hidden, 14, *16*, *17*, *82*
pregnancy
   common secrets about, 14, *16*, *17*, *19*, *82*
   as positive secret, 168–169
prejudice, hidden, 14
preoccupation, 10–12
pre-school children, cognitive development, 35–41
Princeton University, 135
privacy
   avoiding questions, 115
   romantic relationships, 22–23, 110
   vs. secrecy, 8, 13, 19–23, 109–111
prizes, anticipating, 161–162
profession/goals, as dimension of secrets, 85–87, 94–96
profession/work/school discontent, 14, *16*, *17*, *82*, 95

protection of others (coping strategy 2), 98, 100–102
psychology. *See also individual study topics; individual names of social scientists*
   on covering, 118
   mock interview studies, 4, 123
   moral psychology, 87
   personality traits, 24–25, 176
   social psychology, 87–88
punishment, 49–50, 90, 106
   escape from, 71
   self-punishment, 90–91

questions. *See also* concealment
   being asked, 7–8, 36, 108–109, 112
   declining, 112, 114–115
   dodging, 112–115
   thanking for asking, 114–115
   unanswered, 114

reasons for secrets. *See* insight
regret, 70, 143, 145
Reiman, Anna, 122–123
relational mobility, 177–180
relationship, hidden, 14, *16*, *17*, *82*, 94
relationship quality, 50, 109–111, 136
relevance of information
   past relationships, 110
   the right to know, 175
   secrecy *vs.* privacy, 22
religion, 14, 116, 182
R.E.M., 138
revealing secrets. *See* confession; confiding; disclosure
Richman, Laura Smart, 122
romantic desire, 13, 15, *16*, *17*, 18, *82*, 86, 93–94
   crush, 15, 37, 42, 47
romantic discontent, 13, *16*, *17*, *82*, 86, 109

romantic relationships and sex. *See also* sexual orientation
common secrets about, 13, *16, 17,* 18–20, *82*
concealment about, 109–111, 115, 119–120
as dimension of secrets, 85, 92–94
hidden, 14, *16, 17, 82,* 94
infidelity, confessing, 140–141, 145–146
infidelity and culture, 92–93
romantic desire, 93–94
secrecy *vs.* privacy within relationships, 22–23
rumination, 51–55, 59–61, 66–72
causes of, 70–71
consequences of, 71
solutions for, 71–72
Russia, culture and core selves as moral, 78

Saddledome, 164–165
Santa Clara University, 57
sarcasm, 113–114
Savage, Dan, 145–146
Savage Love (Savage, syndicated column), 145–146
savoring, 162–164
school, common secrets about, 14, *16, 17,* 18, *82*
secret joys, 170–172
Secret Telephone, 129–131
secretive, habitual secrecy, 24–25, 50, 58–59, 71, 111
secrets, 3–25. *See also* burden of secrets; children and child development; concealment; confession; confiding; culture; disclosure; positive secrets; secrets on the mind; three dimensions of secrets; *individual common types of secrets; individual study topics*
"big" *vs.* "small," 11

common types, overview, 13–15, *16, 17,* 18–21, *82*
"complete," 19–20, 108
"confided," 19–20, 37, 43–44
country *vs.* common types and personal disposition, 176
defining, 4–8
number of secrets, 17, 20, 24–25, 180
personality and, 24–25, 176
secrecy *vs.* privacy, 8, 13, 19–23, 109–111
shared *vs.* solitary nature of, 6–7
sharing, 189–194
Snowden/NSA example, 4–6, 7, 12, 113
The Sopranos (HBO) example, 3–4, 7, 8, 12, 109
secrets on the mind, 51–73
coping with trauma, 55–59, 152–156
frequency of thoughts, 51–55, 59–61, 66–72
mind-wandering, 62–66
Seinfeld (television show), 88, 160–161
self
authentic, 116–118, 182–183
being true to oneself, 78–80
change, 78–79, 99–100, 135
collectivism *vs.* individualism, 181–186
comfort to be oneself, 120–121
-consciousness, 42
different parts of ourselves, 117–118, 120, 136
-expression, 46, 49, 116, 120–121, 185
individually oriented secrets, 86, 93–94
narrative, 46, 79
self-disclosure, 179–180
self-knowledge, 134–136

self (*cont.*)
  self-monitoring for concealment,
    115–121
  self-punishment, 89–92
  true, 78, 116
  viewing self as moral, 75–78
  at work, 116
self-harm, 13, 16, 17, 77, 82, 94
sex, secrets about. *See* romantic
  relationships and sex
  avoided conversation topic, 47,
    110
  common secrets about, 14, 15, 16,
    17, 18–20, 58–59, 82
  not having sex, 14, 16, 17, 82
  secrecy *vs.* privacy, 21–22
sexual infidelity, 13, 15, 16, 17, 18–19,
    82, 86, 92–93, 146. *See also*
    infidelity
sexual orientation
  common secrets about, 14, 16, 17,
    19, 82
  concealment about, 108–109, 119,
    120, 123–124
  Love, Simon (film) on, 45, 50, 57
shame
  concealing secrets due to, 57–59
  coping compass for, 96–104
  culture, 182–183
  vs. guilt, 100
  self-punishment for, 89–92
Shared Reality (Higgins), 133–134
shooting example, 61
siblings, 43
Singapore, culture and core selves as
  moral, 78
Smart, Nadine, 51–55
Snowden, Edward, 4–6, 7, 12, 102–104,
  113
social closeness and connection, 21,
  94, 136, 147–149, 179
social discontent, 13, 16, 17, 82
social network, 46, 93, 178

social support. *See also* confession;
  confiding; culture; disclosure
  concealment of secrets and,
    110–111, 120–121
  confessing secrets and trust, 143
  confiding secrets, 152
  conveying trust, 140, 151
  culture and emotional expression,
    186–187
  culture and support seeking,
    187–188
  emotional support, 57–58, 71, 155
  perceived concealment, suspicions,
    110–111
  social approval and teenagers, 50
  for stress, 186–187
  struggles, 57
Soprano, Tony, 3, 7, 8, 12, 109
Southern Methodist University, 55
Spain, morality of infidelity, 92
"The Star-Spangled Banner" study, 138
Sticky Note Secrets, 130
Stigma (Goffman), 116
stigmatized, marginalized identity,
  116–118
  identity concealment, 118–124
  situations that reduce belonging,
    117–118
"stop and smell the roses," 162–163
stress
  concealment, 127
  and health, 118
  managing, 154, 169, 186–187
  speech study (positive anticipation),
    160
Strohminger, Nina, 76
Stumbling on Happiness (Gilbert),
  161–162
subway examples
  man on platform story, 74–75
  for mapping dimensions of secrets,
    80–82, 81, 82, 158
  Subway Therapy project, 129–130

suffering, 90, 131, 189
suicide, 55, 64
surprise
    common category of secret, 14, *16,
        17,* 19
    emotional experience of, 165
    gifts, 164–170
    marriage proposals, 164–165, 167
    as positive secrets, 164–170
    risks, 167
    visit example, 166–167

Tamir, Diana, 135–136
TD Bank survey (2017), 95
teenagers, cognitive and social
        development of, 45–50
    parental control, 47, 49, 71
terminal illness secret example,
        173–176, 181, 183–184
thanking others, avoiding questions
        by, 114–115
theft, 13, 15, *16, 17, 82*
therapist, 150
thirty-eight common categories of
        secrets, 13–15, *16, 17,* 18–21,
        24, 59–61, 107, 125, 148, 151,
        175–176
This American Life (NPR), 174, 184
thought suppression, 68–70
    success of, 69
    suppressing familiar thoughts, 69
    suppressing thoughts of secrets,
        69–70
    white bear study, 68–69
thoughts
    communicating difficult thoughts,
        70–71
    duration of thoughts, 64
    extra-relational thoughts, 13, *16, 17,*
        18–20, *82*
    future-focus, 154, 159–162
    number of thoughts, 63
    past-focus, 70, 72

rumination, 51–55, 59–61, 66–72
    thought detection (sensitivity), 66
3D map of secrets, 81–85
three dimensions of secrets, 74–104
    collectivism and, 185–186
    coping compass for, 96–104
    dimension, defined, 79–80
    dimensions of secrets,
        multidimensional scaling, 83–87
    morality and character, 75–78
    morality as dimension of secrets,
        87–92
    morality and perception, 78–80,
        92–93
    professional/goal-orientation
        dimension, 85–87, 94–96
    relationship dimension, 92–94
    subway example, 74–75
toothbrushes study, 39–41
topics of conversation, 23, 47,
        108–110, 112–115, 118, 134
    abrupt shifts, 113
    commonly avoided (money), 22
    commonly avoided (sex and
        relationships), 21–23, 47
    non sequiturs, 113
    postponement, 115
    the self, 134
    small talk, 118
trauma
    coping with, 55–59, 152–156. *See
        also* journaling
    keeping secret, 13, *16, 17, 82, 87,* 96
trolley problem, 87–88
true self, 78, 116
    and good/bad behaviors study, 78
    and secrets study, 77
trust
    avoiding conversations, 110, 114
    being vulnerable, disclosing, 140,
        151, 179
    damaged or destroyed, 143
    fears of rejection and, 50

trust (*cont.*)
  perceiving concealment, cycles of
    concealment, 111
  the risk of secrecy, 145
  and social sharing, 49
  of strangers, 178
truth
  expectations of, 102
  owing others the, 175–176, 182–183
  whether to reveal, 142
Tufts University, 196
Turkey, culture and collectivism, 182

uncertainty, 45, 50, 73, 97, 161
United States
  culture and core selves as moral, 78
  culture and secrecy, 182, 186–188
  morality of infidelity, 92
University of California, Berkeley, 117
University of California, Santa Barbara,
  186
University of Chicago, 83, 89, 138, 141
University of Exeter, 122
University of Melbourne, 90
University of Minnesota, Morris, 63
University of Notre Dame, 58
University of Pennsylvania, 76
University of Virginia, 122, 161
University of Washington, 123
unusual behavior. *See*
  counternormative behavior
urination accidents by children, 35–36

values, culture and, 180–184
violation of trust, 13, 15, *16*, *17*, 20, 82,
  86, 94
vulnerability, 111, 140, 148, 151, 179

Wang, Lulu, 173–176, 181, 183–184
wanting to talk, 124–126
Warren, Frank, 155
watermelon toy study, 32–33
weekend trips, anticipating, 159–160
Wegner, Dan, 122
well-being (mental and physical). *See
  also* burden of secrets
  concealment about secrets and,
    120
  coping with trauma for, 55–59,
    152–156
  mental health, concealing,
    108–109
  secretive and habitual use of
    secrecy, 25, 58–59
  three dimensions of secrets and, 82,
    87–97
  understanding toll of secrets on, 4
"What will people think?," 136–140.
  See also perception
whistleblowing (Snowden/NSA), 4–6,
  7, 12, 102–104, 113
white bear study. *See* thought
  suppression
"white lies," 142
Why We Talk (Dessalles), 133
Wilson, Andrew, 51–55
work. *See* cheating (work or school);
  profession/work/school
  discontent, profession/goals
worry, 48, 50, 72–73, 77, 112, 122, 149
Would they want to know?, 146
wrapping paper, 165
wrongful conviction example,
  51–55, 61

ABOUT THE AUTHOR

MICHAEL SLEPIAN is the Sanford C. Bernstein & Co. Associate Professor of Leadership and Ethics at Columbia University. A recipient of the Rising Star Award from the Association for Psychological Science, he is the leading expert on the psychology of secrets. Slepian has authored more than fifty articles on secrecy, truth, and deception. His research has been covered by *The New York Times, The Atlantic, The New Yorker, The Economist, The Wall Street Journal*, the BBC, NPR, and more.

ABOUT THE TYPE

This book was set in Albertina, a typeface created by Dutch calligrapher and designer Chris Brand (1921–98). Brand's original drawings, based on calligraphic principles, were modified considerably to conform to the technological limitations of typesetting in the early 1960s. The development of digital technology later allowed Frank E. Blokland (b. 1959) of the Dutch Type Library to restore the typeface to its creator's original intentions.